BANGOR

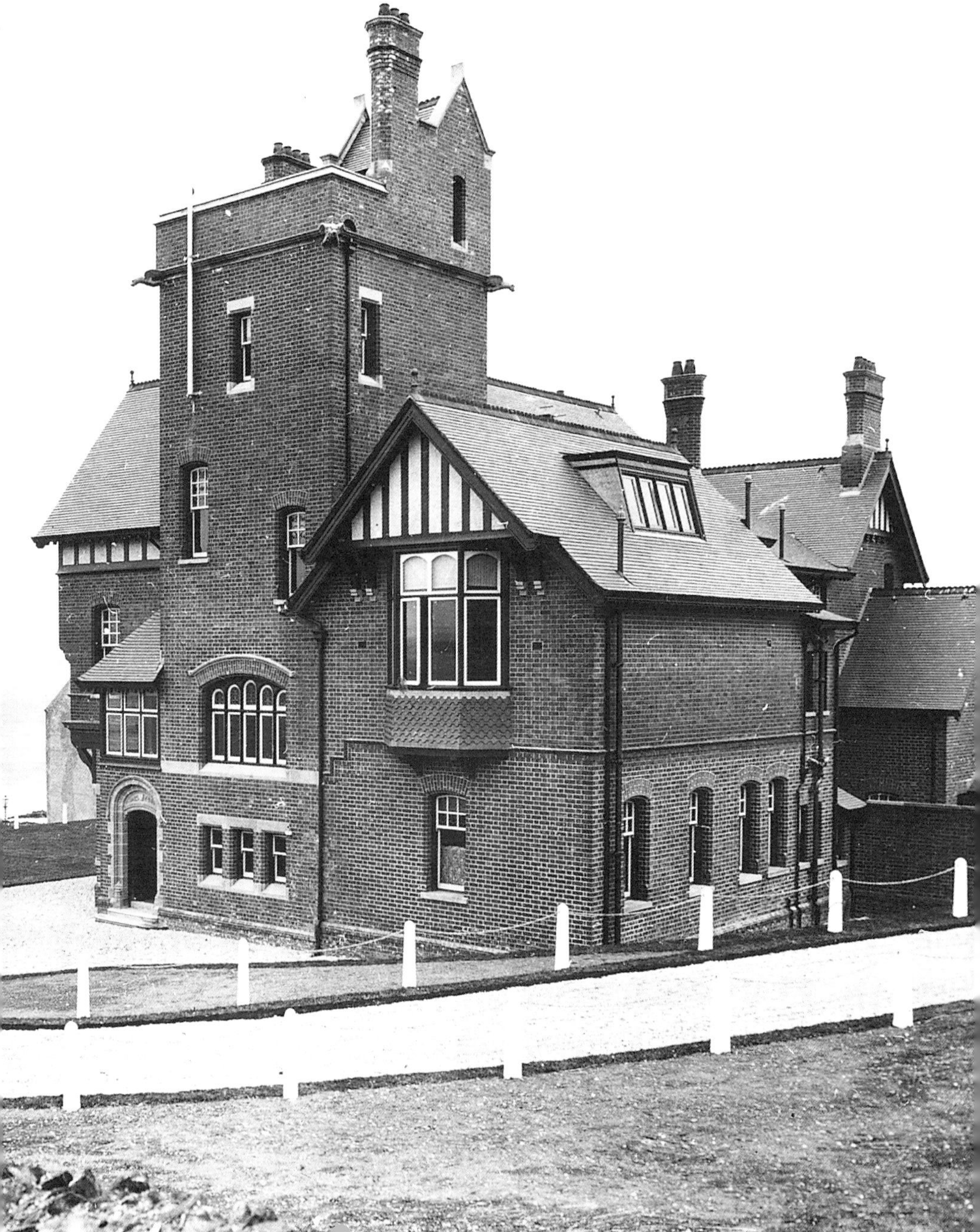

BANGOR
AN HISTORICAL GAZETTEER

including
CARNALEA, CONLIG, THE COPELAND ISLANDS,
CRAWFORDSBURN, GROOMSPORT AND SIX ROAD ENDS

Marcus Patton

Ulster Architectural Heritage Society
1999

© Marcus Patton 1999
Published by the Ulster Architectural Heritage Society
66 Donegall Pass
Belfast BT7 1BU

Typesetting - Sans Souci Gazette
Printing - W & G Baird, Antrim
ISBN: 0 900457 52 X (softback)
       0 900457 53 8 (hardback)

Frontispiece:	*The Royal Ulster Yacht Club photographed shortly after its completion.* (Welch Collection).
Title page:	*Main Street, Bangor, about 1910: when men wore boaters and court houses had cupolas.* (Lawrence Collection).
Tailpieces:	*Dutch barn at 360 Belfast Road.* (Peter O. Marlow).
	Chalet at Donaghadee Road, Groomsport. (Peter O. Marlow).
	The original Golf Club, Hamilton Road. (Irish Builder).
Front cover:	*Queen's Parade ("The Kinnegar") in 1860, as painted by Thomas Hanna.* (NDHC).
Back cover:	*Bangor Abbey.* (Marcus Patton).
	The Coates Memorial. (Marcus Patton).

Acknowledgements

The Ulster Architectural Heritage Society gratefully acknowledges the generous financial assistance towards the publication of this book of Bangor Historical Society, Dunlop Homes Ltd, the Esme Mitchell Trust, Feherty Travel Ltd, T J A Kingan, Knox & Markwell, North Down Borough Council, the Northern Ireland Tourist Board and Mr and Mrs E B Wilson.

A great many people assisted in the preparation of the first edition of this gazetteer, which was then called a list; in particular Ian Wilson and Kenneth Robinson of North Down Borough Council's Heritage Centre, Hugh Dixon, Peter Marlow, Mrs R J Bleakley and Mrs Shelagh Boucher; Denis Gailey, the late Karl Smyth and other members of Bangor Historical Society; my father H A Patton, who had built up extensive photographic records during his time as planning consultant to Bangor Council 1961-72; numerous building owners who kindly allowed me to explore their houses; the staff of the Planning Office in Downpatrick who permitted me to consult bye-law drawings held there, and the staff of what was then the Historic Monuments and Buildings Branch of the Department of the Environment who permitted me to consult their own records of buildings; the staff of the Public Record Office in Belfast and of the Linenhall and Central Libraries; and Peter Rankin, then the overall editor of the Society's publications.

For this revision I have called again on the services of many of the above, and to their names must add Karen Latimer as the Society's current editor, who has been unfailingly prompt and meticulous in her comments, and once again Peter Marlow, who has provided many of the photographs for this edition; both have given great encouragement and support. Terence Reeves-Smyth at the Buildings Record provided useful material on gardens and monuments in particular. The Co Down Spectator is now available on microfilm and I was able to consult it at Bangor Library, assisted by the excellent index compiled by Jack McCoy. Sadly the bye-law drawings, which were such a valuable source fifteen years ago, seem to have gone AWOL from the Downpatrick Planning Office during a relocation some years ago.

In addition, thanks are due to the following for the use of illustrations: North Down Borough Council Heritage Centre (pages x, 87, 110, 127, 171, 177, 178); the National Library of Ireland for photographs in the Lawrence Collection (pages iii, 5, 26, 73, 196); the Trustees of the National Museums and Galleries of Northern Ireland for photographs in the Welch and Hogg Collections (frontispiece and pages 5, 15, 34, 47, 68, 87, 133, 158); Gordon and Donald Finlay (page 99), H A Patton (Pages 26, 41, 105, 106, 118, 177); John Gilbert (page 124); and Tony Merrick (page 9). All other photographs are by Peter O. Marlow. The maps on pages xi, xvi and on the endpage are reproduced from Ordnance Survey maps with kind permission.

A NOTE ON FORMAT

This gazetteer is an extensively revised and enlarged version of the original List of *Historic Buildings in Bangor and Groomsport* published by the UAHS in 1984, and covers the complete urban area of Bangor, including Ballyholme, Carnalea, Rathgael and Primacy, with the addition of the outlying but related villages of Conlig, Crawfordsburn, Groomsport and the Copeland Islands. These areas and their street names are incorporated in the alphabetical sequence of the gazetteer, but cross-referenced from each village name.

As in other recent UAHS lists, the streets are arranged alphabetically, with odd-numbered buildings given first. Unless otherwise stated, all streets are in Bangor itself. **Bold** headings commence entries on current buildings and streets; if a building name in bold follows the number it indicates that it is the original (or at least a very early) name. SMALL CAPITALS are used to identify descriptions of buildings formerly on a site or streets that have been cleared, and *italics* to indicate alternative names, referrals to other entries, and organisations previously using the building or site. Many Bangor buildings originally had names rather than numbers, but cross-references are only given for terraces of three or more houses. Numbers in the margins indicate the pages on which buildings are illustrated. A number of stop press additions made in 1999 are recorded in [square brackets]. *Large italics* are used for former names of terraces that are now known by their street numbers.

Bangor has a great many good late Victorian and Edwardian houses of standard designs - typically two-storey stucco buildings with canted bay windows, Bangor slate roofs (that's the Welsh Bangor, by the way), and red clay ridges - and it has not been possible to mention many of these individually. Their omission, however, does not reflect any lack of appreciation of their contribution to the overall quality of Bangor's urban environment, and sympathetic retention of their sash windows, panel doors and stucco details is well worth while.

The main streets are described comprehensively, and inclusion of a building does not necessarily imply approval of its design or condition - as the ensuing text will often make clear. This is a portrait of Bangor in 1998, warts and all.

Introduction

"Of late years it has rapidly increased in extent, and many new villas attest to its growing favour as a place of residence with business men of Belfast".[1] Messrs Baddeley and Baxter, authors of Nelson's *Thorough Guide to Ireland* in 1909, could safely be plagiarised in a present-day guide to Bangor; indeed, the total invisibility of Bangor's history as a centre of European learning in the Dark Ages tends to render their additional comment that "the through tourist... will not find much to detain him in the town" as true today as it was then. Bangor, however, had a second period of greatness, when its importance was less international, but when a very comfortable and attractive seaside resort was created for the enjoyment of the middle classes of the north of Ireland: a resort whose domestic architecture is still largely intact in a glorious sweep of villas round the coast from Seacourt at one end of Bangor Bay to Glenganagh at the far end of Ballyholme Bay.

The commercial area of Bangor has sadly changed very much for the worse over the three decades. This is partly as a result of four devastating car bombs but also through a steady process of commercialisation of the town-centre residential property, and in recent years the overdevelopment of some of the large villa gardens. Although this list is primarily a record of the Bangor that exists in 1998, some account is given of buildings that have been lost, to add historical depth to the account and also to emphasise how much of late Victorian Bangor does still remain. Let us hope that the tourist of the 21st century does not find all these late Victorian glories gone like their monastic predecessors and still "not find much to detain him in the town" - plenty remains at present.

Bangor was originally called *Inver Beg*, the Beg being the stream that flowed past the Abbey; in monastic times it was sometimes called the Valley of Angels; but the name *Beanchor* which has evolved into Bangor comes from the Irish words *beanna* and *chor* meaning peaks and curve, presumably referring to Bangor Bay. (Though alternative etymologies derive it from *beanna* meaning a cow's horns, which apparently have a legendary connection with the area, *bannagher* meaning "pointed rocks", and an alternative translation of *beannchor* as "white choir").

The early history of settlement around Bangor is not well documented, apart from a description[2] of an earthwork at Rathgael extending over two acres which has now disappeared - the burning of Rathgualla in the year 618 is recorded in the Annals of the Four Masters. It is likely, however, that there were settlements round Bangor and Ballyholme Bays before the establishment by St Comgall about the year 558 of an abbey of Regular Canons which was to become one of the most important monastic settlements of Early Christian Europe. The school associated with the monastery was so celebrated that when Alfred founded the university of Oxford he is said to have sent to Bangor

INTRODUCTION

for professors!³

Such a renowned and wealthy establishment was an open invitation to Viking raiders. By the ninth century⁴ Vikings had settled in some parts of Ireland - there was a Viking burial on the raised beach at Ballyholme - and in 822 raiders murdered the abbot along with some 900 monks.

In 1125 and 1139 St Malachy rebuilt the abbey and it continued to flourish for some time. By 1469 it had again fallen into a ruinous state and at the Dissolution in 1572 when most of its lands were granted to, or seized by, the O'Neills, the abbey became vacant.

In 1571 Sir Thomas Smith, principal Secretary of State to Queen Elizabeth, was granted a patent "to obtain and govern" the "rich and pleasant country... called Clandeboy". However after an abortive invasion in 1572 which cost Smith £10,000 and the life of his son, he "was merely tricked out of it by the knavery of a Scot, one Hamilton (who was once a schoolmaster, though afterwards made a person of honour)".⁵

Hamilton came to Ireland along with Hugh Montgomery, another Scots adventurer, who "foreseeing that Ireland must be the stage to act upon, it being unsettled... concluded to push for fortunes in that kingdom", and applied to James I for half of Con O'Neill's lands when Con was imprisoned. When Con's name was cleared through Montgomery's good offices, he was granted joint control of Con's kingdom, "under conditions that the land should be planted with British Protestants, and that no grant of fee farm should be made to any persons of meer Irish extraction".⁶ However, the King was persuaded by Sir James Fullerton to grant a one-third share of the lands to Mr James Hamilton since "the vast territories" were "too large for two men of their degree". Accordingly "the whole great Ardes" was split in November 1605 between Montgomery and Hamilton, "that the sea coast might be possessed by Scottish men who would be traders".⁷

Thus Con returned home to a mouth-watering feast of "beeves, colpaghs, sheep, hens, bonny blabber, rusan butter... greddan meal strowans, with snush and bolean", but the two Scots set about the plantation of North Down, Montgomery being based at Newtownards and Hamilton at Bangor.

Sir James Hamilton, afterwards created Viscount Claneboye, was accompanied by Scots from Dunlop in Ayrshire who created the modern town of Bangor. By 1611 the Commissioners visiting Bangor were able to report that the town consisted of "80 newe houses, all inhabited with Scotyshmen and Englishmen", the latter including "20 artificers, who are making materialles of tymber, bricke and stone for another house there" in addition to a "fayre stone house" already built. In 1612 the town was incorporated by Charter and became entitled to send "two understanding and fit men" to any Irish parliament.

In addition to the market and fair at Bangor established in the 1605 patents, King James in 1620 granted Hamilton "that from henceforth for ever there

INTRODUCTION

may and shall be a maritime port" at Bangor, along with a court and prison.[8]

In 1625 Hamilton commissioned Thomas Raven to prepare maps of his lands at Clandeboye, and we have our first map of the town. It shows an informal layout, with two principal streets (the present lower Main Street and High Street), development along the shore linking them, and a third street parallel with Main Street. The church, castle, and a mill can be identified, and a rabbit warren occupies the Kinnegar. The custom house, which was built in 1637, indicates that the port was developing by then, and a small pier was built about 1757.

By 1740, one Michael Echlin wrote that some houses were "built with stone and ruff-casted, not built with Mudd like the rest of Bangor houses",[9] and the town continued to thrive, partly thanks to the continued interest of Hamilton's descendants, the Wards. In 1808, Louisa Ward wrote that "The improvement of the Town of Bangor is the Col's hobby horse".[10]

In 1837, when the population was about 3000, Lewis reported 563 houses, albeit "most... are indifferently built... the streets are neither paved nor lighted, but are kept very clean; and the inhabitants are but indifferently supplied with water".[11] The OS Memoirs at the same date are more explicit: "The cottages are principally built of stone, mostly thatched but in many cases slated. With a few exceptions they are but one storey high. Glass windows are in all cases employed, and a tolerable degree of cleanliness and neatness may be seen to prevail in some instances". The town proper is described as "narrow and straggling... With the exception of a group at the southern end of the town, and a few others scattered through it, the houses are slated. They are built of stone and plastered, most of them whitewashed or otherwise coloured".[12]

Cotton manufacturing was introduced to the town in 1783 by one George Hannay. It was noted in the Ordnance Survey Memoirs that although "From some of the inland points, the high church spire and Bangor Castle form the most conspicuous objects, and as the plantation of the latter almost conceals the town, it gives an aspect really picturesque", but from other points, "the two large cotton factories with their smoky chimneys form the most striking features, and the whole place has a manufacturing, crowded, and dirty appearance".[13] Hannay established a factory of eight buildings about 1800 in partnership with a fellow Scot called McWilliam, in the lower High Street area, and in 1806 they built the New Mill, which was five storeys high, on the edge of the present marina. By 1837 the mills employed nearly 300 people, but they closed following two severe fires in 1856. The Valuation Survey of 1860 records the New Mill as "dilapidated", while the Old Mill had become a store.[14]

The Irish Corporation Commissioners reported on Bangor in 1834 and found 2741 people in the town living in 507 houses, whilst a further 54 houses were vacant, and two more were being built. The town was managed by a Corporation composed of members of the Ward family which maintained the

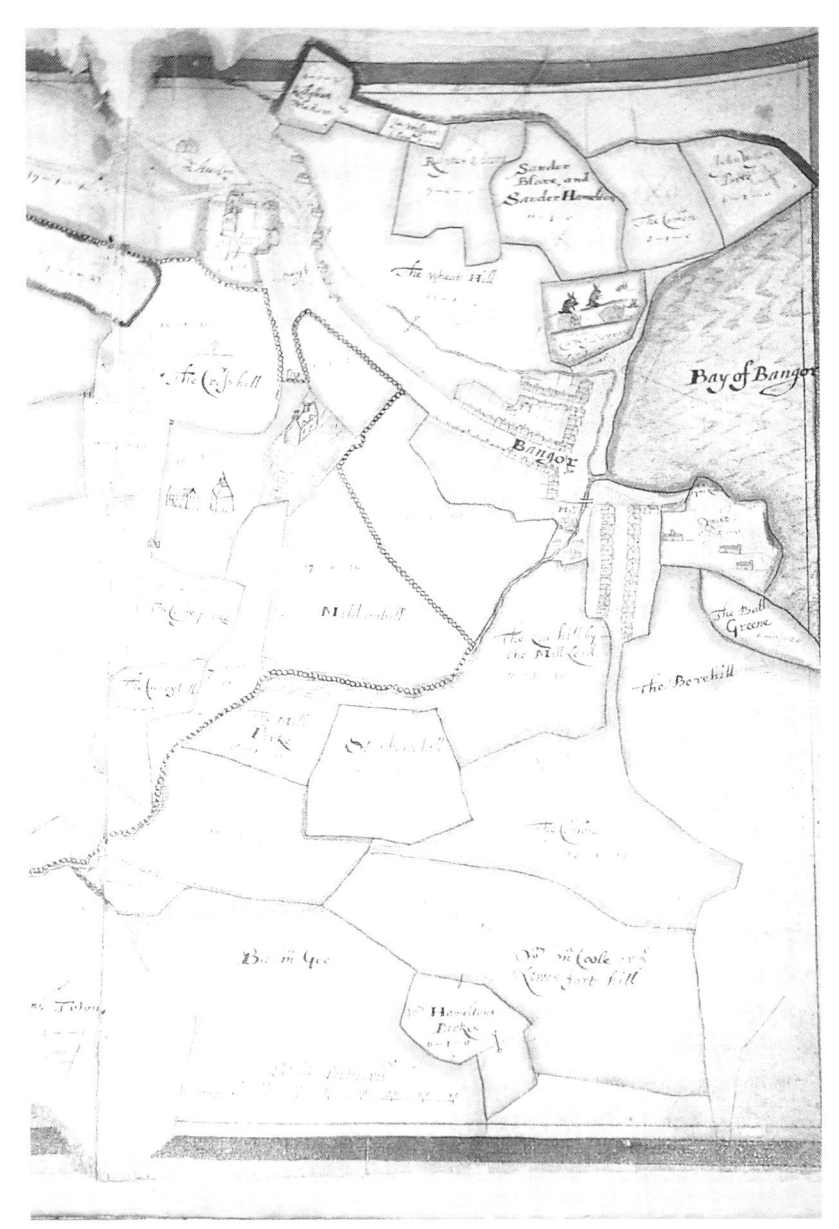

Raven's map of Bangor in 1625: the future Main Street and High Street are already present in embryo, as is part of Queen's Parade. The name of the Kinnegar to the north derives from the rabbit warren shown here. The church, the castle and a mill can also be identified.

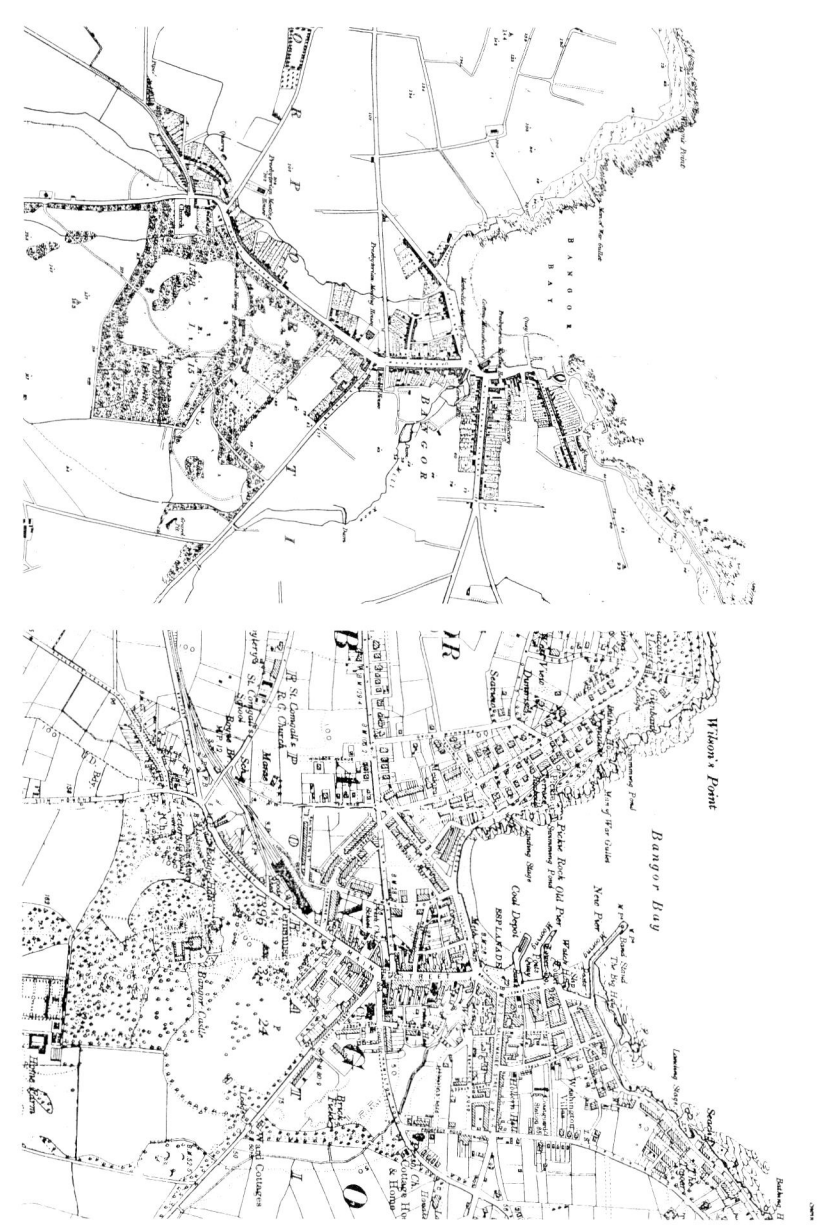

The core of Bangor as seen on the Ordnance Survey maps of 1833 (above) when the cotton mills dominated the seafront, and 1901 (below), when there are three piers and Princetown is well developed.

INTRODUCTION

pier, the market house, the town sergeant (whose salary was 8 guineas a year), a schoolhouse, "a poor-house and mendicity institution" and a savings bank. The Commissioners were impressed by the work of the Corporation as "a rare instance of a property preserved with care, and an income... usefully expended".[15]

In addition to the achievements of the Corporation, Bangor by the 1830s boasted a public library, an historical society, corn and flax mills, three windmills for grinding corn, two Presbyterian meeting-houses, both Primitive and Wesleyan Methodist meeting-houses, several independent and national schools in the area, a dispensary, and even a Sunday school - that formed at Rath-Gael in 1788 by J Rose Cleland being the first in Ireland.

Although the Town Improvement Act passed in 1854 changed things in many towns, the Wards remained in control of Bangor and the Town Commissioners, who were elected in 1862 and ran the town's affairs till 1899 when the borough became an Urban District Council, did not bring about any drastic change of policy. In 1865 wells were sunk in Ballymagee Street (now High Street), Holborn Avenue and Main Street, and reservoirs were built on the outskirts of the town in 1881 and 1891, while main drainage was provided for the town in 1882, and electricity arrived in 1930. An independent gas company founded in 1854 was later municipalised.

Despite continuous improvements of this kind initiated by the Commissioners, there is no doubt that the greatest single influence on the growth of Bangor during the 19th century was the arrival in 1865 of the Belfast, Holywood and Bangor Railway Company. Belfast merchants had since 1848 been able to spend week-ends and summer weeks at Holywood without the inconvenience of being distant from Belfast, but it was only with the arrival of the railway in the town that "that pretty and convenient watering-place"[16] really began to develop as a resort and to become the town we know today.

The *Northern Brighton*, as Bangor had become known, could be reached by road, rail or steamer - the *Erin* and the *Bangor Castle* plying the Bangor-Belfast run thrice daily from the mid-1860s, with both ships apparently managed by teetotal captains and crews. By 1885 "the chastely designed, elegant villas" were "overspreading the landscape, and occupying hill and vale and points of vantage on every side", rising "tier upon tier in varied architectural style and crowning the heights which, on the land side encompass it like the arc of an amphitheatre". During "the Bangor Season" it was estimated that the floating population of the town totalled "something over NINE THOUSAND SOULS".[17]

The Railway opened on 18th May 1865 and, in order to drum up a more regular business than the seasonal summer trade, the Company offered building tickets or "house free" tickets: "for each house of the annual value of £25 PLV a free 1st class ticket; and for each house of the value of £18 a free 2nd class ticket, to and from Belfast... such ticket to attach to the house

INTRODUCTION

for ten years from the date of completion" provided it was sited within a mile of the station.[18] Tickets for boat and train were interchangeable, and this provided an immense stimulus to building at the western end of Bangor, which was followed towards the end of the century by the development of Ballyholme, where there was a long sandy beach.

Bangor's development in the 19th century was very much based on the sea, whether for business or for pleasure. By 1886 there were some thirty sea captains resident in Bangor, mostly plying the route between Bangor and Belfast, and although it never became the "Scotch packet station", the series of piers built at Bangor bear witness to a successful trade in coal and other commodities. However it was the pleasure boats that dominated Bangor including the J-class and 12-metre yachts sailing from the Royal Ulster Yacht Club, the smaller boats that congregated at the Bangor Sailing Club (later to become Ballyholme Yacht Club) and Lenaghan's (later Laird's) rowing boats at Pickie. "In former days", Lord Dufferin told members of the RUYC in 1896, "there was scarcely a pleasure sail to be seen on the lough, while now it is populous with every description of vessel, from the tiny canoe and the half-raters to the ocean cruiser".

Then there was the swimming: even in 1837 Lewis could report that Bangor was "much frequented for sea-bathing during the summer", and swimming matches were organised in the 1860s at the Pickie Rock. Ballyholme at that time was "quite neglected", but the *Irish Builder* reported on "rugged places in the rocks marked 'Ladies Rock' and 'Gentleman's Rock'. These, with the addition of a promiscuous shed and a plank jutting out over the water, along which naked men run in sight of gods and men, are all the facilities afforded."[19] Bathing houses were erected shortly after at Pickie and Clifton for men, and at Skippingstone for ladies, while Pickie Pool was constructed in 1931.

As the resort developed, the town created tourist facilities to attract summer visitors and attempts were made to beautify the town by various improvements. The ruins of the "new" mill and some cottages were removed in the 1890s to form the Esplanade at the foot of High Street, and a bandstand was provided to this promenade; in 1905 the seafront round to Stricklands Glen was acquired by Act of Parliament, and in 1907 Ballyholme Park and in 1909 Ward Park were acquired, followed by Stricklands Glen itself. Commercial enterprises also catered for tourists, from the Grand Hotel built to dominate the Promenade in the 1890s to the refaced Savoy of the 1930s which ushered Bangor into the jazz age; and for entertainment Bangor was provided with the largest cinema in Ireland (and one of the most spectacular) in the Tonic. With two golf clubs, tennis courts, Caproni's ice cream, putting, bowls, bathing, and a Palais de Danse, Bangor was indeed an ideal holiday place in the inter-war years, and it was no wonder that Ballyholme beach was "alive from morn till eve in the season".[20]

Since the last war the linen embroidery industry which thrived in the early part of the century has disappeared, and latterly even the tourist industry has

INTRODUCTION

declined with the growing popularity of foreign holidays, but Bangor has continued to grow - from 3,006 in 1881 through 14,000 in 1930 to 56,200 in the urban area of the town in 1996. Charlie Seyers, who recorded the history of Bangor in a series of entertaining articles in the Co Down Spectator in the 1930s, remarked that the town was controlled by house agents, all of whom "seem to be able to get butter on their bread and some of them manage to get it on both sides."[21] In 1934, it was the healthy climate of Bangor that sold "these labour-saving houses, soundly constructed on modern lines", as recommended by Alexander Devon.[22] The ease of commuting by car, and then the outflow of population from Belfast in the early days of the Troubles to what was seen as the safety of suburbia, led to a vast increase in Bangor's population. This was for a while contained within the Ring Road of the early 1970s, but has now hopped over even that barrier to swallow up the outlying hamlets like Primacy. Bangor is now the second-largest urban area in the province, after Belfast, and the bungalows and Housing Trust estates that marked the town boundary in the 1960s have been succeeded by further waves of housing, now aggrandised with "Georgian" porches or "luxury executive" double garages.

As indicated at the beginning of this introduction, the centre of the town became much more commercialised in the 1960s and this led to the loss of several good properties; that trend was succeeded by the growth of out-of-town shopping centres in the 1980s, which have left many of the earlier commercial developments declining or vacant. Perhaps more worrying, and more insidious, is the frighteningly rapid spread of replacement windows. In the 1970s, night-vent windows, aluminium double glazing and Kentucky doors replaced the often perfectly sound Victorian sashes and panel doors, and in the last decade they in turn have been replaced with the currently fashionable plastic. People often do not appreciate how much of the character of their house depends on its doors and windows, and on features like a natural slate roof and the intricate stucco mouldings that grace many Bangor buildings.

At the time this list was first published in 1984, Bangor had just lost its former Grand Hotel on the seafront, and that heralded a decade of decline while the marina was developed in Bangor Bay. The marina was intended to attract investment, and indeed developers did move in once the property market had declined sufficiently. Several of the fine villas on the Marine Esplanade have been demolished and their sites redeveloped more intensively, while the grounds of Glenbank and Seacourt, two of the finest villas which should have been preserved intact in their original setting, are now packed with new houses jostling against their aristocratic neighbours. In 1992, Bangor lost its most important Art Deco building, the Tonic Cinema, the front portion of which could surely have been incorporated in the sheltered housing complex that now occupies its site. The film-maker John T Davis, a nephew of the Tonic's architect, said that the cinema's fate "seems typical of the inevitable Philistinism of the wealthy seaside resort of Bangor, where the spending of millions of pounds on a yachting marina takes precedence over preservation of its cultural past."[23]

INTRODUCTION

With a town centre that is no longer thriving it is tempting to look to outside businesses or tourism to revive its fortunes, and the Borough Council has entered into discussions with developers to put a conference centre on Queen's Parade, and to replace the thriving market with another shopping centre. Yet surely what Bangor really needs is more people living in the heart of the town, who will use the shops in the main streets every day and who will not want to drive to inconvenient out-of-town shopping centres? Apartments over shops along Queen's Parade, some in refurbished old buildings, others in new ones on the gap sites, would be the most effective way of halting Bangor's present decline. People living in an area want it to thrive and be attractive, and will work to make it that way. Visitors to a town and the residents of its suburbs have no such commitment: a renewed pride in the historic centre of Bangor may be the key to its future.

In assembling this gazetteer, I have tried to look objectively at the recent buildings of Bangor, but it is hard to find many that emulate the splendour and pride of their Victorian predecessors. Not all the recent alterations to older buildings are reversible, unfortunately, but it is to be hoped that the owners of unaltered buildings will continue to look after them, and that new owners will seek to undo at least some of the philistine ravages of recent years.

[1] Baddeley p.54
[2] O'Laverty II pp 130-31
[3] Lewis I p.181, and other sources
[4] Arch Surv pp.102, 140
[5] Life of Sir Thomas Smith (1698), qu. Lowry pp.25-8
[6] Lowry pp.10, 15
[7] Ibid p.16
[8] Ibid pp.xix-xxviii
[9] Stevenson p.49
[10] PRONI D2032.1/10/115
[11] Lewis I p.181
[12] OS Mems pp.23-24
[13] Ibid p.24
[14] Morton p.19
[15] Lowry p.lxxxvi
[16] BNL 19 May 1865
[17] Lyttle pp.30-32
[18] Minutes of Belfast & Co Down Railway, qu. Morton p.28
[19] IB 15 July 1867
[20] BT Guide p.113
[21] Spectator Nov 1932
[22] BNL 27 Oct 1934
[23] Spectator 6 April 1995

The Princetown area of Bangor as recorded by the Ordnance Survey in 1919: the spacious and leisurely development of the Victorian and Edwardian eras had come to an end, and future development would be much tighter.

AN HISTORICAL GAZETTEER OF BANGOR

A

ABBEY

Bangor Abbey Parish Church: 15th century and later: Although largely 19th century in date, this church is the most tangible reminder of Bangor's distinguished early history as a monastic settlement, which had close links with Iona, and from whence monks such as St Columbanus and St Gall travelled across Europe. Such was the fame of the Abbey that Bangor is one of only four places in Ireland marked on the celebrated mediaeval map of the world, the "Mappa Mundi" in Hereford Cathedral.

Of St Comgall's monastic foundation in 558 AD nothing remains, although the shaft of an 8th century Celtic cross found here is kept at Clandeboye Chapel (it may have stood on the Cross Hill indicated on the Raven Map), a sundial of similar age can be seen in front of Bangor Castle, and the Bangor Antiphonary written in Bangor 680-691 is preserved in Milan. Despite its international standing, Bangor's 7th century monastery probably consisted of groups of wattle huts, some communal buildings in timber and a small church of dry-stone, the whole surrounded by a vallum, or rampart and ditch. A succession of Viking raids in the 8th and 9th centuries destroyed the early buildings, notably in 822 when Comgall's shrine was plundered and the Abbot and his "learned men and bishops were smitten with the sword."

The subsequent decline of the monastery (due also in part to the Brehon system of hereditary abbots) was reversed when St Malachy was appointed Abbot of Bangor and Bishop of Connor in 1124. He wasted no time, and "within a few days there was an oratory, or church, finished of timber pieces made smooth, fitly and firmly knit together". This monastery was destroyed however in 1127 by Conor O'Loughlin, and Malachy and his monks left Bangor till 1137. In 1139 Malachy was on his travels again, and after his return from the Continent he "thought it good to have a church built of stone (*oratorium lapideum*) proportioned like to those he had seen in other countries. And when he had begun to lay the foundation thereof, the native inhabitants of the country began to make a wonder thereat, because there were not found in that land as yet such manner of buildings (*quod in terra illa necdum ejusmodi aedificia invenirentur*)." Since a "stone church" (*Daimhliag*) is mentioned at Bangor in the *Annals of the Four Masters* under 1065, St Malachy's Augustinian monastery appears to have been revolutionary in its use of lime mortar (or possibly, as O'Laverty suggests, in its magnificence) rather than in its use of stone. It is possibly a part of this building which is Bangor's oldest surviving architectural fragment - a rubble **wall**, incorporating some dressed stones, stands behind the nearby gate-lodge, and was described by Harris as "a small Part of the Ruins of Malachy's Building" which "yet subsists".

Before his death in 1148, Malachy had established "a noble institution, inhabited by many thousands of monks, the head of many monasteries, a place truly sanctified, and so fruitful in saints" yet by 1469, partly as a result of the Statute of Kilkenny in 1367 which ruled that no "mere Irishman" could make his profession in a religious house among the English - so dispossessing the Celtic monks who had carried Christianity through the dark ages - the abbey had fallen into ruin. In that year Nicholas O'Hegarty took over as abbot and seems to have rebuilt the church. The present tower dates probably from this time, though it was later heightened when the steeple was added. Decay set in again after the Dissolution of the Monasteries in 1542, and the church was burnt by Sir Brian McPhelim O'Neill in 1572. The ruins were still impressive about 1610 when Fleming wrote in *Collectanea Sacra* that there were still "some structures, and vast walls of white stone, and various enclosures, all of which betoken its former grandeur."

When Sir James Hamilton came to Bangor he rebuilt the church as a parish church "within the old abbey about the year 1616", according to Harris. The work was directed by a master-mason called William Stennors, and completed about 1623. He incorporated the 15th century tower (where Con O'Neill had hidden on his escape from Carrickfergus in 1598) and may well have used other stones from the old buildings, since in 1643 Father McCann in his *Irish Itinerary* describes where "stood the monastery of Benchor, once the most celebrated in the whole world, of which even the ruins do not now exist." Hamilton's church was attached to the east of the old tower, without either chancel or transepts - the invaluable Raven shows it clearly. To it an octagonal spire was added, as recorded by inscriptions inside the tower: "This steepel was raised anno 1693 Io Blackwood Io Cleland Church wardens". In 1777, the Hibernian Magazine described the church as "a pretty good building with a tolerable steeple", and said that the PARSONAGE adjoining it was "a handsome new building, and by much the best house in the town".

Lewis wrote that in the course of enlargements to the Church in 1832-33 the foundation of the church "was so much disturbed by injudicious excavations that it was found necessary to take it down, with the exception of the tower; and a spacious and handsome structure, in the later style of English architecture, was erected in the following year, at an expense of £935". (The OS Memoirs were less impressed, describing the church as "having nothing very attractive or ornamental in its appearance"). In 1844 the chancel and transepts were added, the south one containing the old family vault of Sir James Hamilton. As Bangor grew, the congregation needed more spacious premises, and the Abbey church was forsaken in 1882 for the larger and more central new Parish Church of St Comgall (see *Castle Street*). Such was the pace of Bangor's growth however that the Abbey was brought back into use as a second church in 1917 by Rev J A Carey, and in 1941 the Parish of Bangor Abbey (as opposed to Bangor) was created. A new hall was built at a cost of £7,000 to designs by Samuel McIlveen and opened in September 1951. In the course of further renovations in 1960, the ceiling over the crossing was raised, the floor of the South transept was lowered, the organ moved to the back of the church and the East window replaced by a mural. The former RECTORY, described variously as a "good glebe house" and "a great gaunt grey building", was built about 1850, presumably on the site of the 18th century parsonage, but demolished in 1931 and later replaced by the present hall. "Just one more of the old landmarks that are wiped out", as Charlie Seyers observed.

The church is entered through a pointed barrel vault under the 15th century tower, whose walls are six feet thick; above the entrance can be seen an arch which would originally have been open to the West end of the nave but was filled in in the 17th century by the present door and window. The first floor of the tower is reached by an external staircase of 17th century date, and the second floor is probably contemporary with the spire. The tower is capped by a plain balustrade with corner finials, and the octagonal spire rises with broaches from a square base. Recent renovations involved unfortunate ribbon-pointing of the tower, but the remainder of the church is simply roughcast and externally has little interest. Its general style is Victorian Gothic in the transepts, the chancel being expressed by a lower ridge line. Internally, once through the base of the tower, the church is light and simple, with ribbed ceiling vaults rising from trefoil-fretted spandrels. Two good stained glass windows (one to James Steele Nicholson of 1899 by J Clarke & Son of Dublin and one to Richard Ivan Robson of 1917 by Ward & Partners of Belfast) which originally formed lancets flanking the chancel and facing the main body of the church have now become sidelights for the chancel, becoming rather lost in the process.

The monuments are the great glory of the present church, the finest being that to James Hamilton, carried out in 1760 by Peter Scheemaker, with a fine marble statue and cameo-busts of Hamilton and his wife Sophia Mordaunt, situated at the base of the tower. William Stennors the master-mason died in 1626 and his memorial, ornamented with the tools of his trade, is also there. In the church yard are many good 18th and 19th century slate headstones with beautiful naïve lettering; many stones commemorate deaths at sea, notably the splendidly ornamented stone to Captain George Colville who after a brief career during which he "dauntless trod ye fluctuating sea", was

ABBEY

drowned in the wreck of the *Amazon* at Ballyholme in 1780 (a cannon salvaged from the wreck is placed outside the church wall). Other memorials of interest belong to W G Lyttle the journalist and author (1896), George Hannay who established the cotton mills (1821), Archibel Wilson, a young stone mason who was hanged at his home town of Conlig for his part in the 1798 rebellion, and John Simpson, the surgeon aboard the Titanic. There is a story (no doubt apocryphal) told of a drunk falling asleep in the churchyard amidst this illustrious company, and being disturbed in the morning by a factory hooter. Waking up and seeing where he was, he concluded that he had heard Gabriel's trumpet. "Boys-oh-boys!" he said, "Not a soul risen but me. This speaks bad for Bangor!"

See *Adamson; Arch Survey pp 265-6, pls.106,111; BHS I pp.52-54, II pp.6-9; Camblin p.3; Eakin; Hamilton; Handbook; Harris; IB 22 Dec 1917; Lawrence 9550, 9551; Lewis I p.183; Merrick p.4 et seq; O'Laverty II p.126-27; OS Mems p.24; Reeves p.362; Robinson p.89; Scott passim; Seyers pp.22-26; Spectator 29 Sep 1951; UJA 2nd series vol.VI pp.191-204, vol.VII pp.18-36 and vol.VIII pp.173-5; Welch 32-34.*

ABBEY STREET

From Main Street to Brunswick Road. A mostly one-sided street of low two-storey houses facing Castle Park, with most of the houses late 19th century in date, but even so not one in its original state, and most now turned into shops in very poor taste. The building line remains consistent, as does the general scale of the street, but not much else can be said for it architecturally.

In the early 19th century this area was known as *Church Quarter*, and consisted of mostly thatched cottages (described rather grandly in the Parliamentary Gazetteer as "partly unedified, partly one-sided, partly two-sided, partly sub-ramified and aggregately possessing the character of a suburb") which looked on the demesne wall which at that time ran round Bangor Castle. The area included a *Mendicity Institute* which distributed food and clothing to the poor, and a *Municipal Pound* on the site of the present Brunswick Manor flats. There were two SCHOOLS in the Church Quarter in the 1830s.

See *Montgomery p.lxxxix; OS Mems pp.23-24; Parl Gaz p.214-15.*

34 **Gate Lodge:** 1852, probably by Anthony Salvin: Built as a gate lodge to Bangor Castle, this carries the monogram of Robert Edward Ward over its Gothic doorway. One-and-a-half storey ashlar stone lodge in picturesque Jacobethan style with ornamental stone skews and finials, with stone labels over horizontally-divided trefoil-cusped windows. The Castle demesne wall which was demolished *c.*1950 would have made the function of the lodge clearer than it now is. Immediately behind it is the fragment of St Malachy's 12th-century stone CHURCH (see Abbey), and alongside it stood the 19th century SCHOOL HOUSE.
See *Dean p.64.*

Bangor Abbey about 1900: on the site of St Comgall's 6th century monastery, the 15th century tower and late 17th century steeple rise over the early 19th century nave and transepts, with a rather Gothic foreground. (Welch Collection).

Bangor Railway Station as it left the hands of Lanyon Lynn & Lanyon, with polychrome brickwork, attached tower and arcades of round-headed windows. The building is now rendered and haphazardly extended. (Lawrence Collection).

ABBEY STREET

5 **Railway Station:** 1865 and 1890 by Lanyon Lynn & Lanyon, refaced *c*.1950 and further altered 1985: If a competition was ever held for the most unrecognisable building by Charles Lanyon, this would surely win it. Begun by the contractors Edwards Bros in 1862, the line from Holywood to Bangor was opened as a single-line track on 18 May 1865, and its terminus was indeed designed by Lanyon's office - possibly by John Lanyon - an Italianate design in polychrome brickwork with stucco arcaded side porch and pyramid-roofed square tower. The *Belfast Holywood and Bangor Railway Company* operated eight trains daily each way, and contributed greatly to the wealth and size of Bangor (see *Introduction*); in 1884 the Company was absorbed by the Belfast and Co Down Railway, and from 1897 the line became double-track. Shortly before this, about 1890, the terminus appears to have been enlarged by the addition of the clock tower and a concourse spanned by a Belfast roof-truss, to which at some stage was added a prominent notice advising passengers "To Make the Most of Beautiful Bangor, Boldly Beware of Betting and Booze". Another of the company's alliterative slogans was the promise to take passengers to "Bangor and Back for a Bob".

Sadly things have changed since that day in May 1865 when "the numerous travellers seemed greatly pleased with the excellent arrangements of the company": the station-master's HOUSE (which stood below the station on Abbey Street) is gone, along with the original roof of the concourse (now replaced by a depressing corrugated metal cladding), the weather vane and roof have been lopped off the tower, and the oculi, blind arcading and groups of round-headed windows along the body of the station were all plastered over when the building was taken over by the Ulster Transport Authority after the last war. Further alterations were carried out in 1985, reducing the former concourse to a narrow passageway and removing Sharman D Neill's clock and the former Belfast trusses over the main concourse.

See *BNL 19 May 1865; Eakin; Hogg 18-24; Lawrence 4723, 9554; McCutcheon pp.138-157, 179, 218; Morton pp.27-28; Patterson pp.5-6, 9; PRONI, UTA 21/5; Spectator 14 Nov 1985.*

Bus station: 1950: A functional red-brick building with exposed concrete lintels, set back to allow buses to turn.
See *Spectator 17 Jun 1950; 22 Jan 1998.*

Ailsa Terrace: See 74-80 Seacliff Road.

ALBERT STREET
Street from High Street to Victoria Road, consisting mostly of terraced late 19th century two-storey houses, the terraces at nos.8-14 and 16-22 being known as *Ruthville Terrace* and *Cliftonville Terrace* respectively. Somewhere here however in 1910 was Mrs Copley's *Pavilion of Varieties*, which showed a different play each night during the summer of 1910, opening with "the sparkling pantomime entitled Robinson Crusoe".

Nos.5-9: *c*.1890: Terrace of three-storey roughcast houses with lined stucco ground floor; doors with sidelights curiously on one side only.

Nos.25-39 and 32-46: *c*.1890: Facing terraces of two-storey stucco houses with dentilled cornice and continuous hood mouldings. Much altered; original sash windows were horizontally divided.

ALFRED STREET
Late 19th and early 20th century development of stucco houses, from High Street crossing Beatrice Street to a cul-de-sac.

Nos.13-25: 1902-04, by Kenneth McWhinney for C Neill: Terrace of two-storey two bay stucco houses with segmental-headed windows in chamfered opes.
See *App 96*.

Housing Executive office (on corner with Beatrice Road): *c*.1985: Two-storey red brick office with first floor slate-clad: an alien and rather intrusive design.

Nos.10-20: Linwood Terrace: Two-storey two bay stucco houses with high wallhead and a gilt-lettered name plaque on no.14.

Ardbraccan Terrace: See 69-77 Clifton Road and Victoria Terrace.

Ardmore Cottages: See 30-40 Princetown Road.

ARRAS PARK
Cul-de-sac of two-storey hipped terrace houses off Gransha Road, grouped in a courtyard arrangement with long front gardens converging on the short entrance road. Developed by 1928.

ASHLEY DRIVE
From Ashley Gardens to Windmill Road.

Ballyholme Presbyterian Church: by E P Lamont, 1959-61: Red rustic brick church with artificial stone dressings, simply-traceried leaded lights, three-storey square-topped bell tower at corner alongside gable front. The builder was McMillan & Jeffers.
See *Spectator 20 Feb 1959, 10 & 17 Feb 1961*.

ASHLEY GARDENS
From Groomsport Road to Ashley Drive, developed about 1930 and mostly developed by 1939. Five people were killed and thirty injured here in the German blitz of April 1941.
See *Blitz p.xi*.

ASHLEY PARK
Road from Groomsport Road to Ashley Drive, laid out during the 1930s.

ASHLEY PARK

No.38: *c.*1935: Two-storey rendered house with madly irregular brick zig-zag quoins, so erratic that they do not quite complete their journey up the building.

ASHLOANEN: see *Brunswick Road.*

Astoria Terrace: See 35-49 Dufferin Avenue.

Auburn Terrace: See 101-105 Groomsport Road.

Ava Terrace: See 23-35 Hamilton Road.

AVA STREET
L-shaped street off Clandeboye Road, of identical two-storey semis, all modernised with picture windows. Laid out and fully developed by 1930.

B

Bachelor's Walk: See Ward Avenue.

Balfour Terrace: See 24-27 Ballyholme Esplanade.

BALLOO ROAD
From Bloomfield Road South to Newtownards Road, a country road that has been heavily developed in the last couple of decades. Sir John Newell Jordan, who was born at Balloo about 1852, went on to become His Majesty's Envoy Extraordinary and Minister Plenipotentiary at Peking in 1917.
See *BHS III pp.17-18.*

No.43: Rathgael House: *c.*1970: The present-day Rathgael House (*cf* the original one on *Rathgael Road*) is a seven-storey curtain-walled office block on a podium, occupied by the Department of Education.

Nos.12-18: Blackwood: *c.*1850 and later: Two-storey house with small-pane double-hung sash windows, extended to the east and rear. The site was occupied before 1833 but the building, known as *Gresham Lodge* in 1932, and *Hayford* in 1939, has undergone considerable changes and extension.

9   BALLOO HOUSE: A ragged clump of beech trees, standing in a mess of factories, industrial estates and tarmac, is all that marks the site of Balloo House, the three-storey 18th century house of the Steele-Nicholson family, which was demolished in 1976 following a fire two years earlier caused by vandalism while it was in local government hands. Dean records a gate lodge that had been demolished by 1858. There was a family mausoleum built into a low mound in the grounds in 1792 by William Nicholson, but this too was

Balloo House: symmetrical and simply elegant, with elliptical-arched doorcase and small-pane windows. The c.1810 portion of the house photographed shortly before its demolition in 1973. (Tony Merrick).

48 Donaghadee Road: a double-fronted Edwardian villa with massive kneelers to the gables. (Peter O Marlow).

2 Downshire Road: one of Ernest Woods' best Edwardian villas; the right hand bay was originally battlemented. (Peter O. Marlow).

demolished in 1976, and a pair of octagonal stucco gate pillars that still stood at the entrance in 1984 have also gone.
See *Arch Surv p.351, pl.156; Dean p.61; Merrick pp.1-2; Seyers p.35.*

BALLYCROCHAN ROAD
Curving road leading south from Donaghadee Road through Ballycroghan townland towards Six Road Ends. In existence since before 1833, when there was a windmill near the junction with the Donaghadee Road. Open countryside till comparatively recently, when it has been developed with lop-sided houses presenting gables and white arcades to the road.

No.61: Millbank House: *c*.1800 and later: In 1833, a FLOUR MILL and CORN MILL operated from a mill dam on this site; the flour mill was converted to flax scutching but closed down before 1900, by which time the occupant of Millbank was concentrating on the corn mill. In 1970 Charlie Munro recorded a "ruinous" wooden mill wheel and other machinery in the mill buildings, along with date stones giving the owners in 1798 as John McConnell and as James Lowry in 1841. The large millstones at the entrance to the modern Millbank housing development presumably came from here.

Ballygilbert: See Belfast Road.

Ballygrainey: See Six Road Ends.

BALLYHOLME ESPLANADE
A group of late 19th century terraces of two- and three-storey houses, mostly grouped in twos and threes, and almost all stucco-faced, running from the end of Ballyholme Road at Waverley Drive along an embankment overlooking Ballyholme beach. There are bridges at each end of the road: at the west end, the redundant *Folly Bridge*, and at the east end an unnamed bridge which still fords a stream. The terrace and road are at the top of a long high bank on the seaward side down to the sandy beach, "the very spot for civilised bathing", as the *Irish Builder* noted in 1867, "but quite neglected". The council built thirty bathing boxes below the promenade in 1914, but determined swimmers always made for the rockier shores further west, leaving Ballyholme to paddlers.

The lands were originally owned by two brothers called Lamont, John Lamont living in a neat white-washed cottage called *The Farm* near Waverley Drive till his death in 1915. Before the sea wall was constructed each winter took its toll of the bank, and Mr Lamont "could clearly remember potatoes being grown on a patch of the sea-shore below the present high-water mark". Although five groups of houses are named in the 1906 Directory, development appears to have been less formal, several of the named "terraces" having been built in stages. There are few houses of individual merit, but the Esplanade is still a very attractive piece of townscape despite the usual

intrusions of modern windows and box dormers. The houses on the shore beyond the bridge are accessed from Groomsport Road (*qv*).
See *Crosbie pp.17-19; Eakin; Hogg 2, 3; Lawrence 4724, 11225, 11226, C6013; IB 15 Jul 1867; Seyers pp.11-12; Spectator 20 Feb 1914, 23 Dec 1915; WAG 344, 392, 3333; Welch 54-57.*

Folly Bridge: pre 1858: A stone retaining wall to the road constructed of basalt and schist rubble, which originally crossed a stream and was at one time called *Holy Bridge* - it has now been largely filled by tarmac and an 11,000 volt electricity sub-station. The stream drains the golf course and runs from the Donaghadee Road round the back of some Fourth Avenue houses and is then culverted underground to this point.
See *Lawrence 11223, 11224, 11225, C6013.*

Nos.1-11: Ballyholme Terrace: *c*.1890: Stucco fronted terrace. Nos.3-6 are three-storey houses with two-storey bow windows and lugged mouldings to openings, while the pair at nos.10-11 (*St Helier's*) is framed by pilasters.
See *Lawrence 11225, C6013.*

Nos.12-21: Belfast Terrace: *c*.1900: Terrace of two-and-a-half to three-storey mostly stucco-fronted houses (nos.12-14 probably had brick upper floors originally) with slate roofs. A few of the original horizontally-divided sash windows survive. Nos.17-21 have two-storey canted bays and elaborate bargeboards to dormers (now missing from two houses).
See *Lawrence 11225.*

Nos.22-23: *c*.1910: Handed pair of houses in terrace, with a glazed common porch, canted bays, and large pitched roof dormers with duple sash windows.

Nos.24-27: Balfour Terrace: *c*.1900: Terrace of three-storey stucco houses with full height canted bays. Some of the two-panel doors, horizontally-divided sash windows and stucco chimney-stacks still survive.

Nos.28-29: *c*.1935: Pair of two-storey pebbledashed semis; leaded lights survive at no.29. This site was formerly occupied by a thatched COTTAGE with its gable to the sea.
See *BHS II p.27.*

Nos.32-42: Victoria Terrace: *c*.1890: Irregular terrace of stucco-fronted houses with slate roofs. Nos.38-40, one of which has acquired the sign from *Primrose Cottage* that used to stand in Dufferin Avenue, are one-and-a-half storey stucco houses with canted bays, panel doors in roll moulded surrounds, and originally simple gablets (two now altered) with paired round-headed windows. The pairs of two-storey three bay houses to either side of the Cottages have moulded openings, and nos.41-42 have barrel-roof dormers.
See *Lawrence 11226.*

Nos.43-58: Bay View Terrace: *c*.1900: Irregular terrace of mostly stucco houses, almost all altered by the addition of box dormers or alterations to windows. Some good doorcases with fruity corbels survive, along with canted bays to many houses. Nos.49-50 were originally polychrome brick houses.

BALLYHOLME ESPLANADE

No.58 has a grander doorcase and appears to have been built separately from the terrace, with no.57 perhaps having been an infill. In 1905 Bayview House, Ballyholme, was advertised for sale as a "double dwelling house" with nine bedrooms, stable for two horses and free railway ticket for six years.
See *Lawrence 11226; Spectator 24 Feb 1905.*

The Bridge: *c.*1780: Random stone bridge over small stream with dressed stone voussoirs to archway and granite stone coping to inland side. A small arrow slit exists on the seaward side but appears to be a later insertion. Although part of the bridge structure has been covered over with tarmac, most of the upper surface of the archway is still exposed. Part of the bridge and sea wall was washed away on the night of the Big Wind, 22 Dec 1894. The sea wall has been extended in both directions from the bridge. McCutcheon suggests that the existence of the bridge is an indication of the use made of coastal routes at a time when much of the inland country was boggy or forested, and certainly this was the main road to Groomsport up till the 1830s. The stream itself rises near Newtownards and changes its name according to the townland it is viewed from, being known here as the *Ballyholme River*, but elsewhere as the *Ballycrochan* or *Cottown River*; at low tide it runs out into a brackish delta of sea-grass. In the 1890s, stone age flints were found here, along with a Viking grave. Further digging in 1968 exposed more Bronze Age items, along with pottery and horse shoes dated to the 13th century. Submerged peat with stumps of fir trees has been located below the present high water mark, an indication of the retreating coast at this point.
See *Arch Surv pp.102, 140, pl.82; McCutcheon p.35, pl.7.1; Milligan p.37; Spectator 9 Jan 1914, 10 Apr 1986; WAG 2705.*

BALLYHOLME ROAD

From Clifton Road to Waverley Drive. A road of solid suburban villas developed in the final years of the last century and the first decades of this one. The mature gardens and slight changes in gradient and direction of the road are very pleasant.
See *Lawrence 11227, C2358.*

Nos.1-3: This site was formerly occupied by MOUNT HERALD, a stucco house built about 1865 for James Skillen. It was occupied in the 1880s by the writer W G Lyttle, and later by building contractor James Savage. It was demolished in March 1982 for road widening.
See *Seyers pp.27-28, 54; Wilson pp.41, 64-65.*

No.7: Bethany: *c.*1905: One-and-a-half and two-storey double-fronted stucco house with fretted bargeboards to steep gables, and Gothic first floor windows above castellated bow windows. Sadly, the windows have been replaced in plastic.

Clifton School: 1961, by D W Boyd: Single-storey rustic brick block set

back from road.
See *IB 30 Sep 1961*.

No.21: *c*.1900, by W Cooper: One-and-a-half storey double-fronted stucco house with fretted bargeboards to gables at front and a curious triangular window over the doorcase.
See *App 33*.

Nos.25-33: *c*.1895: Detached two-storey stucco double-fronted houses, originally unpainted, mostly with segmental-headed windows. Many details altered, but no.31 has an ornamental doorcase.

No.35: Soldini: 1896, by James Savage for Mr Young: Two-storey stucco house with projecting gable at left hand side. Projecting eaves with boarded soffit, windows with moulded surrounds and keystones.

No.37: Coolavin: *c*.1900: Double-fronted stucco house with moulded stucco gable chimneys and green-blue slates. Segmental-headed double-hung sashes with moulded surrounds and simple doorcase surround with keystone.

Nos.39-41 Elmsthorpe: *c*.1900: Pair of red brick semis with sandstone dressings and crested red clay ridge; chimneys set forward of gable.

No.43: Westhorpe, originally *Ingleside*: *c*.1905: Red brick house with sandstone dressings, rich in leaded glass; bargeboards and iron balcony; original gate piers.
See *Lawrence C2358*.

Nos.47-57, 63: *c*.1900: Variations on the double-fronted two-storey stucco house theme, some still with original double-hung sash windows. Several by Henry T Fulton, and no.53 (*Mar Lodge*) by W H Minshull, both Belfast architects. Most were originally named (nos.47-51 respectively *Walmer*, *Craigelwyn*, *Roseneath*, nos.55 and 57 *Marville* and *Ingledene*).
See *Lawrence 11227, C2358*.

No.59: *c*.1900: Asymmetrical detached stucco house with one bay rectangular and gabled, the other canted with hipped roof and terracotta finial. Unusual narrow sash windows with eight-pane upper sashes. **No.61** was similar but now has plastic windows.
See *Lawrence 11227, C2358*.

No.63: Formerly a two-storey double-fronted stucco house of *c*.1900, with two-storey bow windows, which in the 1930s was used as *Coomehurst School*. [Apartment block under construction 1999].

No.65: Lauriston: *c*.1901, by Henry T Fulton: Double-fronted two-storey stucco house with two-storey bow windows, stucco chimneys and label moulding over door.
See *App 63*.

Ballyholme Park: Acquired by Bangor Council in 1907 and laid out as a park, incorporating a timber shelter and two barrel-shaped mounds.
See *Milligan p.37; Spectator 26 Feb 1909*.

BALLYHOLME ROAD

No.83: *c*.1910: One-and-a-half storey white roughcast bungalow with sturdy rustic brick chimneys rising from a mellow green hipped roof, and splendid leaded-glass windows overlooking the sea. This was probably *The Bungalow*, occupied by Thomas Brand of Brand's Arcade in 1920.
See *Welch 55, 56*.

No.85: *c*.1895: Two-storey stucco house facing the sea above a stone wall to the esplanade, with Arts and Crafts windows to front.
See *Lawrence 4724; Hogg 2; Welch 56*.

Nos.87-91: *c*.1900: Three-storey red brick terrace with red sandstone lintels. No.91, slightly larger than its neighbours and with its gable towards Ballymacormick, was the former *Seabreeze Cafe*.
See *Eakin; Lawrence 4724*.

No.8: *c*.1905: Two-storey double-fronted stucco house with two-storey bow windows with finials. Windows segmental-headed with moulded surrounds; first floor balcony supported on decorative columns over leaded glass front door. Probably originally called *Maymount*.

Nos.18 and 20: *c*.1900: Pair of two-storey double-fronted stucco houses both greatly extended to the rear; ornamental bargeboards to no.18; chimneys removed from no.20.

No.22: Fold Mews, originally *Ellendene*, and later *Roslyn*: *c*.1900: Two-storey double-fronted stucco house with corbelled stucco chimneys and cockscomb ridge, considerably extended. The poet John Hewitt spent six months here with his uncle during the First World War, and later recalled its "gardens front and rear, / that at the front in summer sure to please / with scent and colour lavish everywhere, / that round the back half grass, half apple trees." Its once-similar neighbour, *Lindow*, has been even more drastically altered.
See *Hewitt p.10*.

No.26: Cotswold: *c*.1900: Two-storey red brick double-fronted house with cockscomb ridge and frilly eaves boards; leaded glass lights at ground floor and fishscale slate roofs and iron balconettes to single-storey bays.
See *Spectator 24 June 1983*.

No.32: *c*.1900: Double-fronted two-storey stucco house with a cast iron balcony entered from a stained glass door between the two-storey bay windows at front.

No.34: Oakleigh: *c*.1910: Two-storey red brick house with terracotta panels; canted bays at corners, one with pyramidal roof, the other rising to a gable.

Nos.36-38: Holmlea: *c*.1903, by James G Lindsay: Pair of two-and-a-half and three-storey stucco semis sharing a good mass of corbelled chimneys above the party wall; lunette windows to sides.
See *App 54*.

No.40: Ivy Lodge: *c*.1905: Asymmetrical two-storey red brick house with

62 Ballyholme Road: John Russell's villa of 1901 is set off by elaborate eaves boards and has a varied ground floor plan on a raised site overlooking Ballyholme Bay. (Peter O. Marlow).

44 Ballyholme Road: a rural version of Stockbroker's Tudor with steep bellcast gables. (Peter O. Marlow).

Castle Street: Bangor High School shortly before its completion, with the original margin-paned windows. (Hogg Collection).

BALLYHOLME ROAD

first floor roughcast, battered chimneys; timber verandah. **No.42**, *Westlawn*, is stucco and pebbledash, dating from *c*.1906.

15 **No.44: Tudor Lodge:** *c*.1925: Asymmetrical roughcast house with two-storey gable rising alongside low porch overshadowed by a steep bellcast red tiled roof in Disney Tudor idiom.

15 **Nos.48-52: Baypark:** *c*.1870: Three-storey terrace in brick with stucco ground floor; shallow canted bays at ground floor have margin-paned sashes, while the pilastered doorcases carry rosettes on their capitals; red and yellow quarry-tile paths originally ran up to each door, and the gable to First Avenue used to be softened by a group of trees.

No.60: Sandhurst: *c*.1900: Double-fronted stucco villa with the original front door between canted bays now converted to window, and windows converted to plastic.
See *Welch 55, 56.*

No.62: Villa le Bas: dated 1901, by John Russell for David Warnock: Irregular two-storey red brick villa on commanding site with a complex hipped roof set off by ornamental crestings, finials, dormers and tall chimneys, with ornamental brickwork and terracotta panels below.
See *App 41; Welch 55, 56.*

Nos.66-68: Aurora: *c*.1905: Pair of two-storey stucco semi-villas with canted outer bays set in front of escutcheoned gables. Balustrading over central porch.
See *Lawrence 4724.*

Ballyholme Terrace: See 1-11 Ballyholme Esplanade.

BALLYMACONNELL ROAD

From Groomsport Road to Donaghadee Road, interrupted by the East Circular Road. Road present before 1833, with an ancient *fort* at the northern end and a *Schoolhouse* near the junction with the present Towerview. The OS Memoirs record no less than three corn mills, two flax mills and a windmill in the townland of Ballymaconnell in the 1830s.
See *OS Mems p.22.*

Nos.89-93: *c*.1910: Detached hipped-roof stucco rural workers' cottages with quoins to corners and opes. Five-barred gates and iron pillars like field gates here and at the adjacent terrace **nos.81-87**.

St Columbanus Secondary School: 1960: Two-storey brick and glass building.
See *Spectator 21 Aug 1959, 2 Dec 1960.*

BALLYMACORMICK ROAD

From Circular Road eastwards towards Groomsport, one side almost entirely developed since 1960 and the lands to the north, which are currently playing fields, currently under threat of development.

Nos.34-36: Duquesne and Avoca: *c*.1800: Although these houses are of some antiquity, being present on the 1833 OS map, and still thatched, they have unfortunately been considerably altered, with picture windows inserted and chimney stacks rebuilt in brick. Nevertheless they have considerable interest in an area of otherwise bland new development.

BALLYMAGEE STREET: see *High Street*.

BALLYMULLAN ROAD, Crawfordsburn

Road linking Crawfordsburn to the Ballyrobert Road, narrow and without footpaths, with mature gardens creating the feeling of an English country lane. Most houses altered or modern, but no.1 makes a good eye-catcher for the end of the main street, and there are some interesting outbuildings alongside no.3.

BALLYROBERT

The hamlet of Ballyrobert is little more than a punctuation at a junction on the Belfast Road, most of its buildings having been considerably altered. The short terrace comprising *Pax Cottage* and *Jubilee Terrace* could however be made quite attractive with proper restoration of doors and windows, and there is a pretty cottage with a cartwheel fanlight set among trees on the hillside above the settlement on the Belfast side, which is just outside the scope of this list.

Ballyrobert Orange Hall: 1877 with later alterations: Two-storey five bay roughcast building with shallow rectangular bay set forward asymmetrically. The building was enlarged in 1937 and again *c*.1960.

Bangor Abbey: See Abbey.

Bangor Castle: See Castle.

BANGOR ROAD, Conlig

Section of road from the Main Street of Conlig towards the Old Bangor Road.

No.41: Dunoon: *c*.1890: Smooth-rendered house with high wallhead and narrow double-hung sashes; off-centre porch.

Tank: *c*.1935: A startling circular concrete tank some eighty feet in diameter and twelve feet high, which looks like a miltary bomb shelter but is presumably associated with the nearby reservoirs that were laid out by Henry Chappell in the 1890s.
See *IB 15 Jan 1895 p.27*.

BANGOR ROAD, Groomsport

The Old Rectory: *c*.1840 and later: Two-storey three-bay stucco house with

quoins and skewed gables. Later porch obscures elliptical spiderweb fanlight and lozenge sidelights. Distinctive rubble-stone wall to the road, with saddleback top. The early history of the building is confused: it appears to have started life as a Perceval Maxwell house - *Rose Lodge* - built about 1840; the Presbyterian Rev Isaac Mack lived here *c.*1870 and extended or partly rebuilt it, but by 1885 it was used as a Rectory, and had become known as *Albertville*.
See *Nelson pp.24-25*.

No.21: Islet Hill Farm: *c.*1840: Two-storey three-bay stucco house with outbuildings on the site of a former rath.

Glenganagh Farm: pre 1833, but altered: To the NE of Glenganagh, farmhouse and outbuildings.

No.39: Glenganagh: *c.*1820 with later alterations: Two-storey roughcast house with dressed stone opes; label mouldings to casement windows and over pointed-arch door with sidelights; lozenge chimney-stacks grouped in twos and threes. Fine cast-iron conservatory and verandah with barley-sugar downpipes and ornamental spandrels. Whether it was built as a dower house or had an earlier history, the Dowager Lady Dufferin was certainly in residence here in the 1850s; in 1875 it was owned by a Mr Andrew Cowan, and about 1880 Samuel Kingan JP acquired it and "expended vast sums in ornamenting the place"; it was presumably he who added the conservatory "heated on the most approved principles". Although the house is visible from Luke's Point across Ballyholme Bay, it is "the shady copses of Glengannagh", as Praeger described them, which contribute so much to the beauty of this stretch of coast from Ballyholme to Ballymacormick Point.
See *Lyttle p.39.*

Gate Lodge: *c.*1900, by James Hanna: At the entrance to the well-wooded estate of Glenganagh is a gatescreen and hipped-roof gate lodge. The roughcast lodge, which replaces an earlier one, has a Doric doorcase in red sandstone, battered stone angle buttresses and dentilled eaves. This part of the road between Groomsport and Bangor has a wonderful wooded section full of primroses and bluebells.
See *Dean p.77.*

The Manse: 1880-81: Two-storey double-fronted stucco building set among pine trees, with canted bays and apex board over twin round-headed gable windows. Built by Rev James Latimer (contractor James Fletcher of Bangor) for £747, on what was reputed to be the site of a fairy fort (that is, a prehistoric rath).

No.36: Glenganagh Cottage: *c.*1880: Tiny roughcast cottage with porch and horizontally divided windows.

BANK LANE
Narrow lane from Quay Street to Holborn Avenue, named from the former

Belfast Bank at 16 Quay Street. Mostly resurfaced, but the improvers seem to have run out of tarmac just below Albert Street, where there is a section of cambered granite setts.

BANKS LANE
Banks Lane is a narrow road from the Groomsport Road near the Groomsport roundabout to the shore. It is lined with rubble stone walls, partly capped with large flint stones. The name comes from Banks Cottage (see *Groomsport Road*), which was served from the lane.

BAY LANDS
An estate of houses largely laid out and designed by the young architect Gordon O'Neill on a speculative basis during the 1920s and 1930s. "Bay Lands by the sea" was being advertised in 1922 and 1923 as sixteen acres of beautiful building sites to let with roads made and services laid on, "free from dust in the summer and sheltered from the winter storms". While the houses have since become nicely weathered, the mature gardens that should by now be enhancing this Metroland-type development have not materialised - there are too many concrete lawns and too few trees, and the meandering and hilly streets are far bleaker than they need be; far from growing, indeed, many gardens appear to have dwindled, with burgeoning trees removed, hedges uprooted to make wide openings for cars, and some whole gardens lost to tandem development. O'Neill, whose mother was described in 1952 as "Bangor's grand old lady of temperance", later moved to Chelmsford where his practice restored churches.

See individual street entries: *First Avenue*, *Second Avenue*, *Third Avenue*, *Fourth Avenue* and *Sixth Avenue*.
See *Spectator 2 May 1952; Yearbooks 1922 and 1923*.

Bayview: See 2-6 Seacliff Road.

Bay View Terrace: See 43-58 Ballyholme Esplanade.

BEATRICE AVENUE
Short street of plain two-storey two bay houses, apparently built about 1900, from High Street to Beatrice Road. Horizontally-divided sash windows survive at no.14. At one time Bangor had a town crier who would announce auctions and other events, Mr Alex Jamison who lived at no.5. He was a painter to trade, but had worked on sailing ships in his youth; he died in 1945.
See *Spectator 20 Jan 1945*.

BEATRICE ROAD
Terrace of two-storey stucco houses from Prospect Road to Bingham Street, developed at the turn of the century. Sometimes given as *Beatrice Street*.

BEATRICE ROAD

Nos.3-27: *c*.1908: Two-storey stucco houses with moulded openings and keystones.

No.29: *c*.1905: Two-storey double-fronted house with two gablets, slightly grander than its neighbours in the terrace and still having original segmental-headed sashes and four-panel door. Currently painted a lurid dark green with purple reveals!

Nos.6-28: Dunedin Terrace: *c*.1900: Two-and-a-half storey stucco houses with plain bargeboards to dormers.

Nos.30-32: *c*.1900: Pair of two-storey two bay stucco semis, with first floor windows segmental-headed with moulded surrounds, and hood mouldings and bosses to ground floor opes.
See *App 21.*

Beaumont Terrace: See 47-56 Queen's Parade.

BEECHWOOD AVENUE
Street from Oakwood Avenue to Hawthorn Drive. The line of the road was present in 1833, when it led from Church Street to the flax mill near Spring Hill. It had become a cul-de-sac by 1858, but was laid out again for housing about 1932 and developed during the 1930s, mostly in plain two-storey terraces with exposed rafter ends. A more unusual group of stepped terraces at **nos.22-48** has clay pantile roofs.

BEECHWOOD GARDENS
Rather drab angled cul-de-sac off Beechwood Avenue, of two-storey terrace houses, originally with duple sash windows and smooth cement trims to doors with scotia scrolls and roundels. Laid out before 1940.

BELFAST ROAD
The present Belfast Road, from Abbey Street towards Springhill, was developed about 1850, cutting through part of the original Croft Street and replacing an older road that ran from the present Beechwood Avenue. James McClurge, who laid out the road as far as Clandeboye, lived at the Bangor end and was a strict Presbyterian who would not milk his cow on Sundays. The road only began to be developed between the wars. See also *Old Belfast Road.*
See *Seyers p.15.*

BANGOR LODGE: The main entrance to Clandeboye was formerly graced by an irregular gabled brick gate lodge of 1849 by William Burn, with mullioned lattice windows, set into the entrance quadrant. Latterly in use as the post office, it was demolished about 1960 for a road widening scheme, and there is now a plain bungalow nearby in its place.
See *BNL 14 Jul 1960; Dean p.49; Spectator 5 Aug 1960.*

Ballygilbert Presbyterian Church, Belfast Road: a simple barn-plan church of 1842, with plain Classical details, recently and sympathetically extended to the front. (Peter O. Marlow).

Ballygilbert Manse, Belfast Road: a late Victorian house with rectangular windows set in Gothic arches, and a mediaevalesque half-timbered porch. (Peter O. Marlow).

BELFAST ROAD

Clandeboye House: See separate entry.

Red Row: 1867, for Lord Dufferin: Now rendered, this L-shaped group of a dozen two-storey houses was originally of red brick (hence its name) with stone dressings and lancet windows. Windows stone-mullioned, doors lancet headed, roofs slated with bands of fishscale slates. At each end of the main elevation is a taller gable, the corner one being *Reading Rooms* and a Sunday School. Originally, each house had a front garden, and a small plot at the rear for growing vegetables. The woods behind which set off the row are called *Walmer Grove* (Walmer being one of the Cinque Ports of which Lord Dufferin was Warden).

No.360: Coach Hill: Coach Hill Farm is marked to most people by the Dutch barn that confronts travellers to Belfast with the "gospel in a nutshell" painted in large letters on its side. On the gable of a more permanent barn facing the road is the name of the farm in white pebbles set in the render. On the hill up behind it is a large ring of trees, named by Lord Dufferin *Jan Meyen Clump*.

21 **Ballygilbert Presbyterian Church:** 1842 with later extensions: Erected to serve the district of Crawfordsburn, according to Lyttle in the 1880s this simple stucco church attracted "not a few" people who regularly walked from Bangor, and it continues to draw its congregation from some distance, though most now arrive by car. Set on a simple plinth with a flight of steps up to it, the church has a pedimented frontispiece with a belfry, and simple round-headed windows along the sides. In 1988 the nave was enlarged by adding an extra bay towards the road, but the original entrance was replicated and the extension is now almost invisible.
See *Lyttle p.59.*

21 **Ballygilbert Manse:** *c.*1880, probably for Rev John Quartz: Picturesque two-storey house in random rubble with cream brick dressings to opes, and steeply pitched slate roof with alternating courses of plain and fishscale slates; many windows paired; timber porch. Set in attractively wooded grounds.

Belfast Terrace: See 12-21 Ballyholme Esplanade.

Belgravia: See 176-182 Seacliff Road.

BELLEVUE
Road developed from Groomsport Road to Donaghadee Road about 1930.

26 **No.14: Ballyholme Windmill:** *c.*1780: This tapering circular stone windmill stump is now a roughcast three-storey house. In 1835 it had contained three pairs of stones and "machinery on the most improved construction", but by the 1860s it was derelict and was restored by one John McGilton, who apparently ground corn in it for years after. At least two of the stones were apparently six feet in diameter and nearly two feet thick. A photograph in the Lawrence collection taken about 1900 shows the revolving turret and sails

complete but without fabric, with the present aggregation of outbuildings and fairly generous fenestration. Before the area became so built up, the windmill must have been visible from a considerable distance. It was gutted by fire in July 1922 and later used by a scout troop before being converted to its present form as a house.
See *BT 1 May 1982, 2 Feb 1995; Crosbie p.21; Eakin; Green p.52; Lawrence 4724, 9553; OS Mems p.22; Seyers p.12; Spectator 29 May 1937, 8 Aug 1953, 31 Aug 1957; WAG 345, 3855; Wilson p.91.*

Nos.34-36: *c.*1935: Two-storey semis with rustic brick ground floor, bow windows and Gibbsian doorcases with broken pediments in artificial stone.

Belvoir Terrace: See 68-76 Dufferin Avenue.

BEVERLEY HILLS
Steep winding road from Donaghadee Road to the south, backing on to the golf course. The similarity of the name to the home of the film stars is not coincidental, unlike the town of Holywood, Co Down: the road was developed on the line of a lane that originally led to the farm of Samuel Johnson, whose wife Amy was a devotee of the silver screen.

No.45: *c.*1935: Modernist flat-roofed house with corner windows, originally in rustic brick but recently pebbledashed and given plastic windows.

No.20: *c.*1935: Presumably this was also originally a flat-roofed Modernist house, but about twenty years ago, in a disconcertingly mediaeval effect, it was bizarrely "stone"-clad and converted to something like a castle with chimneys and large windows, and is now painted Disneyland pink.

BINGHAM LANE
Narrow lane running from 75 Main Street to Bingham Street, formerly linking to *Mill Row* at the bottom of Main Street. On the site of the present car park stood the former CHOLERA HOUSE, a building of about 1830 in basalt rubble and brick (some of the latter intriguingly stamped 'TICKLE'); further down was Neill's COAL YARD (also now demolished), which incorporated a Belfast roof truss with a span of 30m, possibly the largest span for this type of truss to have been erected in the province. Neill's also had a number of LIME KILNS in the vicinity (see *Mill Row*). The GAS WORKS (see *High Street*), demolished about 1980, stood near the bottom of the lane and dominated the Bangor skyline for many years with its towering retorts.
See *Allen; McCutcheon pl.131.1 (who also took a large collection of photographs of it now in the Monuments Record); UA, Sept 1989 pp.51-52.*

BINGHAM STREET
A once interesting curved street of two-storey houses from Hamilton Road to High Street, varied in heights, angles to the road and grouping, but uniform in having fresh painted stucco finish and labels over windows. Many houses were apparently built of stone quarried near Pickie, and they got their water

BINGHAM STREET

from a public well in the street. The road is now rudely interrupted by a roundabout near the site of the former gasolier, and is seeing increasing commercial development. Bingham was the family name of Lord Clanmorris, who lived at the Castle.
See *photographs in NDHC; Seyers pp.9, 28.*

No.5: *c.*1908: Two-storey end of terrace house in lined stucco with quoins, with windows framed in Baroque scrollwork; crow-stepped lean-to extension at side.

No.9: The Ganges: *c.*1912: Asymmetrical two-storey house with a frilly bargeboard and verandah supported on chunky fretted corbels.
See *App 398.*

No.29: *c.*1890: Two-storey double-fronted three bay stucco house with quoins; continuous label moulding at first floor, terminating in a boss at each end. Currently on the market with nos.31-33 as a development site.

No.2: *c.*1905: One-and-a-half storey three bay stucco house with paired ground floor sashes.

Nos.4-14: Luskinyarrow Terrace: *c.*1900: Terrace of two-and-a-half storey stucco houses with ground floor bays and label moulding to first floor windows; wallhead dormers with finials.

No.16: Tullyhommon, originally called *Millerville*: 1901, by Henry T Fulton for Mrs Magge: Two-storey stucco house with steep gables set forward at front, containing roundels and flanking an oculus over the front door. Sadly lacking chimneys, and with plastic windows.
See *App 36A.*

No.18: *c.*1908: Similar to no.16, but lacks roundels and has frilly bargeboards instead.
See *App 311.*

Nos.20-28: *c.*1912: Two-storey stucco terrace with quite a number of double-hung sashes, in moulded surrounds with small keystones, and canted ground-floor bays.

BLOOMFIELD PLACE

Nos.1-20: *c.*1945: Two cul-de-sacs of single-storey prefabs, set off by a good group of pine trees nearby. Low corrugated metal roofs and tiny chimneys, neatly kept with manicured gardens. The original builders would be surprised to see the houses - intended as emergency housing at the end of the war - still standing, let alone so obviously cared-for.

BLOOMFIELD ROAD

From Gransha Road to South Circular Road, continuing beyond it as *Bloomfield Road South*. Present before 1833, but little developed before the 1930s.
See *Spectator 4 May 1962.*

No.15: *c*.1925: Chalet with red asbestos slates carried out over timber verandah. A semi-detached version, somewhat altered, survives at **nos.7-9**. At **no.41** a plain little house glamorises itself with ribbons of coloured glass in lieu of quoins.

No.71: Croft Community, originally *Demesne House*: *c*.1910: Two-storey pebbledashed house, originally with double gate pillars and flanking walls. House extended by Robinson Patterson Partnership in 1985, and grounds extensively developed with bungalows since.
See *UA Dec 1985*.

Bloomfield House: *c*.1880: Two-storey double-fronted stucco building with paired windows over ground floor bays.

No.4: Bloomfield Lodge: *c*.1935: Two-storey roughcast house with hipped two-storey porch, and hipped roof carried on down at side to incorporate garage. Westmoreland slates with saddleback ridge and tall chimney-stacks; generally casement windows. Leaded upper lights to shallow bow window; Gothic timber gates in stone boundary wall, good wooded garden.

Bloomfield Centre: *c*.1994, by WDR & RT Taggart for Lochinver: Vast single-storey development in red brick with a central entrance in the shape of a glazed bridge with side towers; domes at intervals to light the internal spaces; clear plastic cloches to house wandering shopping tolleys; and the *pièce de résistance*, no less than twelve acres of car parking.
See *UA Oct 1994 pp.48-49*.

Bowman's Cottages: See 2-16 Somerset Avenue.

Boyne Bridge: See Brunswick Road.

BRAEMAR PARK
Small crescent of houses off the north side of Clifton Road, laid out from about 1935 in the grounds of The Tower (see *Clifton Road*).

BRIDGE STREET
Single-sided street facing what used to be Bangor Bay, between High Street and Main Street. It derives its name from a BRIDGE shown on the Raven map of 1625 fording the stream serving a mill between Main Street and High Street. In the 1860s it consisted of one house (now no.25), but it was fully developed by the turn of the century. Although the basic structures have changed little, most of the buildings are now a sorry sight, with particularly inappropriate window alterations to the upper floors of nos.9-19; the oriel windows inserted earlier at nos.5-7 have a rather cheerful character altogether missing from the later alterations.
See *Hogg 38, 39; Lawrence 2363, 12683, 2859, 4727, 4740, C6019; Seyers p.4*.

No.l: Rough Stuff: *c*.1890: Small two-storey two bay rendered house with

Bellevue: Ballyholme Windmill in 1887, when the skeleton of its sails was still a prominent sight in the Ballyholme area. The building was converted to a house about 1930, and is still occupied. (Lawrence Collection).

Bridge Street in 1968: the view from the Sunken Gardens moves from the Dutch gable of no.3 past the jaunty oriels of no.5 to the striped roofs of the Visitor's Café. (H A Patton).

modern shopfront. In 1906 it was the *Criterion Hotel*.
See *Lawrence 2363, C6019*.

No.3: Red/Green: *c*.1915: Three-storey rendered building with Dutch gable surmounted by a pair of stone children who idly survey the passing crowds. The previous building on this site was similar to no.1 and set back considerably from the road; it was Loughrey's shop, which sold the famous Rosie's Lumps, a form of Yellow Man; Rosie's husband Mickey also sold the Lumps in the street.
See *Lawrence 2859, 12683, 4740; Wilson p.84*.

Nos.5-7: Crisp and Dry: *c*.1890: Three-storey two bay stucco building with jaunty oriels inserted at first floor about 1935, perhaps when it opened as the *Jubilee Café*.
See *Lawrence 12683, 12685, 2859, 4727, 5478, C6019*.

Nos.9-11: Allen & Hamilton: *c*.1995: Four-storey gabled rendered building with rather bland detail, painted sea-blue. This was previously a three-storey three bay rendered building of about 1890 similar and linked to nos.13-17, and known as the *Esplanade Hotel*.
See *Lawrence 4727, 4740, 5478, C6019*.

Nos.13-17: *c*.1890: Brasilia: Three-storey three bay building now much altered, although still with its original striped slate roof. Advertised at the turn of the century as the *Visitor's Café*, it was latterly West's maroon-painted ice cream shop, with a curious little projecting kiosk at the front only seven feet wide.
See *Lawrence 4727, 4740, 5478, C6019*.

No.19: McCullough: *c*.1890: Three-storey building entirely altered, with simple modern shopfront.
See *Lawrence 2363, 4727, 4740*.

Nos.21-25: McCullough's, Barnum's: *c*.1860, with later alterations: Two-storey five bay building with central door to upper floors between shop fronts. Formerly *Bridge House*, this was built as a home by Dr Higginson, and converted to shops *c*.1910.
See *Lawrence 2363, 4727, 4740*.

BROADWAY

Winding road from Donaghadee Road out into the countryside, in existence before 1833, but hardly developed before 1900; for much of its length it is built up only on one side, the houses looking out over the grounds of the golf club.

Bangor Golf Club: 1934, by Samuel Stevenson: Sober white-painted two-storey hipped-roof clubhouse with adjacent pavilions. The break with the club's original Edwardian premises in Moira Drive for this £6,000 Modernist building must have been quite a wrench for its more conservative members, but the sale of part of the old course enabled the club to purchase 61 acres on

Williamson's Farm and to get a new course designed by James Braid; Viscount Craigavon laid the foundation stone in September 1934. The 1964 extension cost £20,000.
See *Bangor Golf Club - 75th Anniversary 1903-78; Hogg 9, 10; Spectator 14 Apr, 8 Sept 1934, 21 Feb 1964, 13 Nov 1997.*

Nos.14-28: *c*.1900: Terrace of two-storey rendered houses, single first floor window over window and door below; bulky rendered chimneys.

No.60: Inglenook: *c*.1935: Two-storey house with walls in rustic brick at ground floor and roughcast above; streamlined tile-hung bay windows with steel windows; rosemary-tiled hipped roof.

BROMPTON ROAD
A steep winding road from Downshire Road down to the shore.

No.1: 1960, by H A Patton for Dr Geoffrey Carey: Smooth-rendered felt-roofed house with asymetrical frontage partly clad in weatherboarding. The low profile masks a relatively spacious interior.

No.3: *c*.1965: Bungalow with roof flipped up to form a two-storey weather-clad frontage.

No.10: *c*.1980: Sleek low rendered bungalow with two asymmetrical gables and picture windows. This sits on a commanding site at the top of a grassy slope looking out to sea, which was formerly occupied by the red-brick HOME OF REST AND CRIPPLE'S HOME built by W J W Roome about 1900.
See *Lawrence 4728; PRONI D.1898/1/29.*

BROOKLYN AVENUE
Elbow-shaped street from Groomsport Road to Windmill Road, laid out about 1930 and developed during that decade with standard house types.

Ballyholme Methodist Church: 1935-36, by H A Hobson: Red brick hall with lancet windows behind a rustic brick gable which has a faience string course and copings and perpendicular-style window. Built by Thornbury Bros of Belfast, and opened in June 1936.
See *Haire p.11; Spectator 2 Feb 1935, 20 June 1936.*

BRUNSWICK ROAD
A gently winding road rising over the steeply-humped *Boyne Bridge* across the railway line, its lower reaches dominated by the square tower of St Comgall's chapel. Originally called the *Ash Loanen*, it existed as a lane through open countryside in 1833. Its houses range in date mostly from the 1860s to the 1930s, many with hedges and small trees.
See *Seyers p.14-15.*

Engine Shed: *c*.1855: Rubble-stone building below the bridge from Abbey Street, with blocked-up round-headed windows and a square red brick chimney alongside. On the other side of the bridge was a brick **signal box** (*c*.1890)

with wavy bargeboard and finial.
See *McCutcheon pp.179-80.*

No.19: *c.*1930: Two-storey roughcast house with two-storey advanced gable with rising sun motif over a rosemary-tiled ground floor bay. Modern small-pane windows retaining leaded glass upper lights. This house-type can also be seen at 2 Rugby Avenue.

Community Centre: *c.*1985: Low shallow-pitched building in rustic brick with slight striping; various roof levels and two small turrets break up roofline. On site of ST COMGALL'S HALL, which was a single-storey building with lancet windows, roughcast with stone quoins, built about 1851 and used as a chapel until the present church was built. It was "converted into schools" in 1887, and later used sporadically to provide additional classrooms, but ST COMGALL'S SCHOOL replaced it in 1929, only to be demolished in its turn.
See *O'Laverty, II pp.153-4; Spectator 8 Nov 1984; Wilson p.57.*

St Comgall's RC Church: 1886-90, by Mortimer Thompson; extended 1983: Long barn-type Gothic Revival church with prominent four-storey square campanile tower attached at east side of front elevation. Traceried pointed arched window containing clustered circles on front elevation. Built of dark grey squared rubble stone with pinky sandstone quoins, for Rev Patrick McConvey. Interior simple, with vegetable decorated columns dividing nave from side-aisle, and clustered veluxes over the altar to throw it into prominence. This church replaced the earlier chapel (see previous entry), which was finally demolished in 1984. Before its erection, "Mass was celebrated in an empty house in Ballymagee Street", records O'Laverty, "which was at other times used for itinerant shows."
See *NDH, 22 Feb 1889; O'Laverty, II pp.153-4, IV p.xxxviii; Spectator 30 Jul 1927, 8 Nov 1984.*

No.27: St Comgall's Presbytery: *c.*1900: Wide two-storey double-fronted red brick building with hipped roof and broad two-storey canted bays with hipped leaded roofs; alternating stone and brick in roundheaded doorcase.

Nos.29-41: Elizaville: *c.*1890: Terrace of plain two-storey stucco houses. No.37 retains the original sashes, four-panel door and a sheeted yard door.

Nos.49-53: *c.*1865: Terrace of low one- and two-storey stucco houses, with high wallheads.

Nos.59-65: *c.*1870: Group of two curiously narrow two-storey two bay stucco houses flanking a pair of semi-detached houses of similar design with ground floor canted bays, built on a hill set back from the road. Much altered in recent years by pebbledashing and replacement windows, though the box dormer on no.65 has acquired a trio of bargeboards.

Nos.71-75: Brunswick Terrace: *c.*1880: Stuccoed rubble-stone houses with simple door surrounds, considerably altered. The fragments of the slightly later, mostly demolished, house at no.75 are said to include the remains of a

doorcase from the old mill at the bottom of High Street.
See *Seyers p.5.*

Nos.79-81: *c.*1935: Pair of double-fronted semi-bungalows, each with scalloped tile-hung gables set forward over square bays with stained-glass toplights; castellated chimney pots.

Nos.99-101 and 107-109: *c.*1950: Two-storey snocrete houses with hipped pantile roofs and pictorial stained glass panels alongside front door.

Trinity Nursery School: *c.*1890, altered *c.*1930: Long roughcast hall with bargeboard and dark rustic brick battered chimneys at each gable. In 1833, this was the site of a PRESBYTERIAN MEETING HOUSE, which also included a *National School* by 1858. The present buildings were erected by James Colville.
See *Seyers pp.15, 35.*

Nos.72-74: Chathwurra: *c.*1910: Pair of three-storey stucco semi-detached houses set an angle to the road; with chamfered opes to windows.

Nos.86-88: *c.*1910: Two-storey roughcast semis on brick plinth with half-timbered outer gables flanking central verandah roof; vertically striped brick and roughcast chimney survives at no.88, with tripartite segmental-headed windows.

Brunswick Terrace: See 71-75 Brunswick Road.

BRYANSBURN GARDENS

Double entry off Bryansburn Road with two modern bungalows on an appendix at one end. Laid out before 1903, presumably as a service road to Bryansburn Road houses and never properly developed.

BRYANSBURN ROAD

A road of late-Victorian and Edwardian houses, which rises westwards from Central Avenue and dips over a hill towards the Bangor West railway station, set off by a large tree in the garden of no.26. Until about 1910 it was called the *Crawfordsburn Road*, since it linked what is now Main Street straight to Crawfordsburn, but it gets its present name from what in 1833 was the *Bryans Burn Bridge* at the western end of the present road.
See *Hogg 12-15; Lawrence 9543, 9545, 9546; Seyers p.35.*

Nos.1-9: *c.*1905: Terrace of two-storey stucco houses with frilly bargeboards complementing the facing West End Terrace.

No.11: *c.*1895: Two-storey double-fronted stucco house with ground floor bow windows and frilly bargeboards.

Nos.21-23: *c.*1900: Two-storey double-fronted stucco houses with frilly bargeboard to gables, and ground floor bays. Dentilled cornice at no.21 (*Clonsilla*), but windows altered; plain sash windows and four-panel door at no.23.

BRYANSBURN ROAD

Nos.25-31: *c.*1905: Group of detached double-fronted two-storey stucco houses with full height bow windows. No.25, *Windermere*, still has original plain sash windows in chamfered opes, balcony at the front, and lean-to conservatory. No.27, *Beulah*, has a balustraded balcony, and no.29, *Artona*, a moulded bargeboard; at no.31, *Landour*, the bays are canted.

No.33: *c.*1910: Two-storey double-fronted house with fretted apex pieces to steep gables. Narrow round-headed first floor windows paired over ground floor bays.

Nos.35-37: *c.*1910: Two-storey red brick semi-detached houses with canted bays: no.35 has the original Arts and Crafts windows with leaded glass.

No.45: *c.*1925: Two-storey roughcast house with tripartite Arts and Crafts windows to outer bays; and duple first floor windows over.

Nos.49-51: *c.*1920: Pair of three-bay two-storey semis with roughcast first floor and red brick ground floor, with escutcheons in gables above two-storey canted outer bays. Tall bulgy pots to red brick chimneys.

No.53: *c.*1900: Two-storey double-fronted house with pleasant frilly eaves and bargeboard, unfortunately modernised.

Nos.55-59: Olga Mount: *c.*1900: Terrace of two-storey stucco houses with ogee gutter on dentilled cornice, and various bays. No.61 is matching infill of *c.*1925.
See *Lawrence 9543*.

Nos.63-65: *c.*1902, H T Fulton for Robert McGuire: Two-and-a-half storey stucco semi-detached houses with two-storey canted bays, and segmental-headed opes.
See *App 81; Lawrence 9543*.

No.69: *c.*1910: Two-storey double-fronted stucco house with verandah over ground floor bays. All windows tri- or bi-partite, in segmental-headed opes with small-paned upper lights over casements.

No.71: *c.*1910: Two-storey double-fronted stucco house with crested ridge; door in lancet-headed ope with hoodmoulding and decorative bosses over.

Nos.81-83: *c.*1925: Roughcast two-storey semis with central pair of half-timbered gables feeding into a mutual dragon gargoyle rainwater hopper. Steep bellcast over porches; leaded lights to casement windows.

Bryansburn Bridge: In the dip of the road, sheltered by a large tree, there is a bridge parapet on this side only; the stream runs under the road to become the stream through Stricklands Glen. Just beyond, where there is now a garage, stood old FLAX MILLS in the early 19th century.

Nos.8-10: *c.*1900: Two and a half storey stucco semi-detached houses with crested red clay ridge and tripartite windows at first floor.

Nos.12-14: Wanstell: 1899, Henry Fulton for Matthew Gibson: Three-storey unpainted stucco semi-detached houses with two-storey canted bays.
See *App 13*.

BRYANSBURN ROAD

Nos.16-18: Fairholme: *c*.1900: Three-storey stucco semi-detached houses, similar to nos.12-14 but with side bays and with hipped roof. No.18 re-rendered with loss of mouldings.

No.20: *c*.1900: Two-storey red brick house with Westmoreland slate roof and barrel-roofed porch.

Nos.22-24: Casaeldona and **Haslemere:** *c*.1900: Two-and-a-half storey semi-villas of similar character to no.20, but with mutual half-timbered and roughcast gable.

Nos.28-30: Arklow: *c*.1890: Two-storey semi-villas with square corner bays rising to steep hipped roofs with terracotta finials, bonnet-roofed dormers and good chimney-stacks; cast iron balconies over the front doors. Rendered, and finished with white pebbles set plum-pudding fashion into the plaster at first floor level.

No.32: Liscarney: *c*.1890: Two-storey house in red brick with pebbled first floor as nos.28-30. Deep soffit; bulgy terracotta chimney pots.

Nos.46-48: *c*.1900, by Young and Mackenzie for Joseph Mercer: Two-storey three bay red brick semi-villas with canted outer bays terminating in jettied half-timbered gables, and apex-boards in side gables; porches with finials and frilly bargeboards.
See *App 17*.

No.52: *c*.1910: Irregular pebbledashed house with mullioned windows, scalloped canted bays and a small conservatory, now rebuilt in plastic.

No.76: *c*.1925: Hipped roughcast two-storey house with stained glass casements and recessed front door.

No.78: *c*.1910: Double-fronted two-storey stucco house with segmental-headed moulded opes. Mature garden with pine trees.

Glen Cottage: *c*.1800 and later alterations: Two-storey building, altered but with apparently early walls including projecting stones near the base, and thick random stone gate pillars.

Bryansburn House: *c*.1880: Two-storey five bay stucco house with two-storey canted bays; hipped slate roof with terracotta finials. Windows horizontally-divided sashes.

BRYANSGLEN PARK

No.46: *c*.1890: Two-storey red brick semi-villas with ornamental bargeboards and finials; some tripartite sash windows; outbuildings of earlier date. Until about 1970 these villas stood in open countryside, but they are now surrounded by bungalows.

BURNSIDE: See *Mill Row*.

C

Caproni's: See 244 Seacliff Road.

CARISBROOKE TERRACE
Terrace of houses alongside Victoria Terrace, reached by a lane off Clifton Road, with rolling lawns in front running down towards Seacliff Road.
See *Lawrence C5054; Welch 20, 21.*

Nos.1-6: *c.*1875: Terrace of two-and-a-half storey stucco houses, with two-storey central bays, many with dormers. Good plasterwork internally.

CARNALEA
Carnalea (originally *Carnaleagh*, meaning a grey cairn, or *Carn a'laogh*, the cairn of the champion) is a distinctive suburb of Bangor, with until recently a collection of rather colourful prefabs filling in the areas between a number of substantial Victorian houses. The prefabs have now largely been replaced by more routine bungalows, but most of the big houses remain.

See *Crawfordsburn Road, Killaire Avenue, Killaire Road, Lowry Hill, Station Drive, Station Road* and *Station Square.*

CASTLE
Bangor Castle: 1847-52: The Victorian Jacobethan mansion house of the Ward family, Bangor Castle was built in 1852 on the imposing hilltop site overlooking Bangor Bay which had been occupied by a succession of earlier mansions. *34*

When Sir James Hamilton came to Bangor he required a fortified house, and the Plantation Commissioners reported in 1611 that he had built "a fayre stone house... about 60 foote longe and 22 foote broade", which appears on Raven's map of 1625. In 1637 however the Commissioners reported on further building activity - "The Lord of Clannaboys... is building of [a] goodly house at Bangor, which according to the plat laid and the stable which is in forwardness, will be one of the fairest in the kingdom". The description implies a new structure, but it may simply have been extensive improvements to the old house.

This house was in decay by 1725 (when the tenant wrote to Hamilton's agent Charles Echlin that "there is hardly a room dry if rain comes and your furniture will suffer much more, and this too much spoyled alreddie"), and in 1779 Luckombe describes only a "low moderate structure". The *Post-Chaise Companion* was obviously referring to a new building in 1803 as a "very elegant house", and this must have been Edward Southwell Ward's Gothic castle which was illustrated in Molloy and Proctor's *Belfast Scenery in Thirty Views* of 1832 and in William Greenwood's sketch of January 1838.

Bangor Castle about 1900: no Victorian mansion was quite complete without a good set of chimney stacks, as seen in this photograph of about 1900; it would be splendid to see the now-missing stacks restored to this otherwise intact building. (Welch Collection).

Abbey Street: the surviving gate lodge to the Castle, probably by Anthony Salvin and dating from about 1852. Originally the Castle was surrounded by a demesne wall, with this guarding the main entrance. (Peter O. Marlow).

Lewis in 1837 describes Bangor Castle as "late the seat of the Rt Hon Col Ward, surrounded with extensive grounds tastefully laid out" - Ward had died that year. His son, Robert Edward Ward, seems not to have shared his father's taste, and he demolished the Castle; according to one account this followed a fire, according to another he simply disliked the draughty structure. (A directory of 1852 records that "The old castle... is still standing, but is to be pulled down when Mr Ward moves into the new one".) A tunnel uncovered in 1985 by workmen at the Leisure Centre, brick-lined and running for some 140m, its floor yielding old bottles and sea shells, appeared to belong to one of the earlier castles.

Ward travelled widely in England, and no less than three architects had a hand in this replacement building. First off the mark in 1845 was one "William Walker, architect", who produced a series of plans and elevations for a Jacobethan mansion with a turreted gatetower, labels over mullioned windows and stone gables above each corner. In a letter to Arabella Ward in March 1845, W R Ward wrote that "Mr Walker... is regularly established in Bangor working away with his plans", and was then "occupied with the stables" with which R E Ward appeared happy; however at any mention of the plans of the main house Ward "very soon dropped it like a hot potato, saying he did not like to venture himself into the contemplation of it, it was so vast a task". Walker was a Monaghan architect, but also appears in Bangor directories for 1846 and 1856. His designs were not in fact built, but Ward seems to have been pleased with the general conception, since the present Castle is not very dissimilar. However, in 1847 he approached William Burn, who had built Dartrey, Co Monaghan, for Lord Dartrey in 1845 and was to prepare designs for Elizabethanising Ballyleidy House (ie Clandeboye House) in 1848. Burn's name appeared as *architectus* along with the builders (rather grandly credited as *opifex*), James Paton and (the stone masons) John Parker of Ayr, on the foundation stone of the Castle dated 28 March 1848, and his obituary credits him with "1847, Bangor Castle Robert E Ward Esq New House and Offices £9000".

That would sound conclusive evidence for Burn being the architect, were it not that the Directory for 1852 gives the cost of the building, which had just been finished, as £20,000, and the date 1852 appears over the porch; and that Anthony Salvin, our third claimant, was credited in *his* obituary, prepared by his clerk, with a new house or substantial alteration for Robert Edward Ward, at a cost of £9,000. Unlike Burn, Salvin is only known to have carried out one other commission in Ireland, the Royal Cork Yacht Club, but both architects were capable of working in the Jacobethan style, and on the basis of this evidence David Walker's suggestion that Burn may have started the building and Salvin finished it seems quite likely. The more mediaeval and serious style and greyer stone of the stable block compared with the main house may confirm the suggestion, but without further documentary evidence we may never know. (As an aside, Mr Parker's two sons, John and Archibald,

CASTLE

who carried out most of the stonework of the Castle, married the two daughters of their Kinnegar landlord, Thomas Whannell, and all four emigrated to Australia, where John became Mayor of Melbourne, and Thomas an MP and Premier of Queensland).

Despite its name, the present Bangor Castle is a very placid mansion house, its picturesque two-storey elevations topped with gablets, strapwork crestings and (mostly now removed) tall Tudor chimneys, all in a warm buff sandstone imported from Ayrshire, with scalloped quoinstones and stone mullions breaking the generally horizontal emphasis of the heavy plinth and string courses. Apart from the vigorous stone fretwork topping each bay window on other elevations, there is more sculpture on the east (entrance) elevation, which is the most irregular, with a three-storey tower at one side capped by an off-centre clocktower and weather vane. The Ward arms with their motto *Sub cruce salus* tops the NE bay window, while at the base of the remaining chimney is the device of a bird under a crown and the motto *Spes Christus mea*, motto of the Clanmorris family (Lord Clanmorris married R E Ward's daughter in 1878). Over the porch door is a plaque reading "Erected by Robert Ward 1852", with more shields and monograms (MC and CC) nearby, while Ward's monogram also appears on rainwater hoppers.

The stables, laundry and service courtyard to the south are in a Gothic style, with slit windows, buttresses, and an octagonal castellated turret at the SW corner. The laundry has been imaginatively adapted to form the *North Down Heritage Centre*, opened October 1984 (Architects: McAdam Design). The extensive demesne is finely planted, and as early as 1744 Harris described the gardens as "large and handsome, and filled with noble Ever Greens of a great Size, cut in various shapes". There are four Irish yews in the formal garden.

Inside, the Castle has survived the change from family mansion to *Town Hall* and Council offices remarkably intact. The Hall is oak-panelled, leading to the Dining Room and Drawing Room with their ornate plasterwork and overmantels, and to the Great Hall or Music Room, the chief room of the house, where Ward could indulge his passion for organ music in mediaeval splendour, lit by the enormous stained glass heraldic window and surrounded by portraits of his family. (One of the heraldic shields in the bottom right-hand corner of the window is rather fetching - Sophia Whaley is represented by three small pink whales.) The organ, a large one with three manuals and pedals by M A Gern of London, appears to have been sold by the Council on the grounds that it would be too expensive to restore, and its present whereabouts is unknown. The first floor consisted of bedrooms (each originally with a tin bath painted in a colour that matched the hangings of the bedroom), the basement of kitchens and servants' quarters, and the attic was the children's domain.

In the grounds of the Castle (formerly surrounded by a demesne wall of stones said to have been quarried at the Long Hole in Seacliff Road), were two **gate**

lodges, one at the entrance to *Castle Park Road* which has been recently demolished; and one on *Abbey Street*, which bears Ward's monogram and the date 1852. The **Home Farm** is now reached from Valentine Road (qv). The family **burial ground** occupies a plot in a wooded area of the grounds. In front of the Castle is the stump of an early carved stone **sun-dial**. By family tradition the stone was used to anchor the Duke of Schomberg's boat in Ballyholme Bay, while a nearby tree stump was reputedly used by Schomberg to tie his horse on his way to the Battle of the Boyne. In the rose garden is a "chapiter brought from Africa by James Hamilton Ward, Admiral", possibly removed from an Early Christian church in North Africa.

On the death of Lady Clanmorris, R E Ward's daughter and heiress, in 1941, much of the Castle grounds was sold to Bangor Council to become Castle Park, and the demesne wall was taken down. (The 7th Baron Clanmorris, who worked for MI5 during the war, wrote detective stories under his real name of John Bingham, and was the model for John le Carré's spymaster Smiley). In 1952 the building itself became the Town Hall for Bangor Borough Council. Apart from removing most of the great Jacobethan chimney-stacks, the Council's custodianship of the Castle has been exemplary; but unfortunately the same cannot be said of the demesne - municipal and educational buildings have greatly reduced the area of the original parkland, and have hardly enhanced it. It is to be hoped that no further development of the park will be permitted.

See *Arch Surv p.228; BHS I pp.52-54; Bingham p.49; BNL 31 Mar 1848; Eakin; information from M C Perveval Price; Ir Arch Archive, Down 6; Lawrence 12686, 12687; Lewis I p.182; Merrick p.196; Molloy; PRONI D1529/2/8; Raven; Seyers pp.15, 29; Spectator 21 June 1906, 16 April and 10 May 1952, 19 Sept 1985; Stevenson pp.49, 285, 288; TRIBA 28 Mar 1870 p.126, qu.Girouard; UJA Oct 1901 pp.161-65; Welch 41-45; Wilson pp.47-48; Young p.199.*

Castlemount Terrace: See 27-67 Castle Street.

CASTLE PARK ROAD

It was perhaps inevitable that once a road had been driven through the grounds of Bangor Castle it would be developed; it was understandable that the Council should seek sites for necessary facilities to improve education in the borough; but it is very unfortunate that so much of the Castle grounds has been sacrificed in the process. Although the main cultivated part of the grounds remains, redolent with eucalyptus scents and packed with rhododendrons and other more unusual plants, the wilder areas on the fringe of this road are being steadily eroded. On the positive side, the retention of large trees along this road makes it very pleasant despite the loss of open space.

Police Station: 1962: Complex of flat-roofed blocks in brown brick, with small square windows, dominated by a large radio mast. When the police moved here from their old barracks in Victoria Road, it was noted that this "well-lighted, cheerfully decorated and most congenial" new station was open

to the lawns of Castle Park; sadly, the security situation of more recent years requires it to be well-screened.
See *Spectator 1 June 1962*.

Leisure Centre: *c*.1973, probably by Hugo Simpson: Complex of brown brick and ribbed concrete, accessed up a concrete ramp.

Glenlola Collegiate School: 1963 by D W Boyd & Co, extended 1972: Sprawling heavily-glazed two-storey steel-framed buildings laid out on the fringes of the park at a cost of £215,000, to which a shanty town of temporary but no doubt nearly as expensive classrooms has been appended. The *Ladies' Collegiate School* had its origins about 1880 as a private school for girls in Upper Clifton. It made several moves, being located at Gray's Hill in the Edwardian period. In 1914 the Misses Weir took the school over and expanded its numbers from 18 to 80, at which point they moved to Pickie Terrace. Despite some expansion on that site, a further move was necessary to Avoca in Princetown Road in 1950, where it amalgamated with *Glenlola School*. The school moved to the present site with 705 pupils.
See *Spectator 3 May 1963*.

GATE LODGE TO BANGOR CASTLE: *c*.1852, probably by Anthony Salvin: Similar to the lodge in Abbey Street but very much altered, with walls rendered and roof tiled, this lodge at the corner of Castle Street and Castle Park Road was demolished in September 1998, to be replaced by a large notice proclaiming itself a "primary announcement" of new apartments. Even its pretty little strapwork gate has gone. [Construction is under way 1999].
See *Dean p.64; Spectator 19 Feb 1998*.

North Down & Ards Institute of Further and Higher Education: 1967, and 1993-94 by Boyd Partnership: The Bangor Technical School was founded in 1900, but soon required better premises, for which the council sought to utilise the funds Andrew Carnegie had offered to assist in the building of a free public library (see *Hamilton Road*). This was extended by the construction of Hamilton House in Hamilton Road which was again outgrown, and the complex moved to the older portion of this building in 1967. The bulk of the latest extension, a red brick building with strong horizontal strips of dark blue glazing, is broken up by a complex zigzag plan; a distinctive detail is the use of triangular projections whose tiled roofs feed rainwater pipes that are carried down the point of the glazing like the nose guards on Norman helmets.
See *Spectator 24 Jul 1993, 23 May 1996*.

CASTLE SQUARE

The square at the end of Market Street was originally a market garden, but about 1880 was laid out with terraced houses for workers on the Ward estate, constructed about 1880 of local red brick with cream dressings, with canted bays on the ground floor. The last of the houses was demolished in December 1984, and the square is now a large car park.

Bangor was granted the right to hold a weekly market in 1605 ("every Monday weekly for ever"). The market which (having been revived in this location in 1923) still operates just off the Square every Wednesday has something of a fair atmosphere with its sales of "Perfect Browns" and "Crackeds" (eggs), flipover buggies, super bubbles, honeycomb, boiled sweets, vegetables, rugs, fish and many cakes baked by church women. After the war, when one of the market sheds had been used as a mortuary, another became the Co-op Hall and it was used as a dance hall every Saturday night, with local jeweller Fred Eakin and his band providing the music. The future of the market, however, is very uncertain, as after several abortive plans for redevelopment in the 1960s, North Down Borough Council has now sold the ground to developers.
See *Montgomery MS pp.i-vii; Seyers p.4; Spectator 18 August 1923, 9 Dec 1960, 8 May 1964, 13 Dec 1984, 27 July 1995, 8 Aug 1996.*

CASTLE STREET

From Main Street to the Gransha Road. Before the development of Hamilton Road at the end of the 19th century, Castle Street had been a narrow two-sided street, but the northern side was taken down to allow the erection of St Comgall's, and the early houses on the southern side gave way to car parks in the sixties. Early businesses in the street had included the chandler Alexander Brown who made all sizes of candles from farthing dips upwards; and John Brown's sewed muslin and embroidery business, which employed up to 900 women and girls from as far as Greyabbey and Ballywalter as home workers, until the American Civil War cut his supply of cotton.
See *Seyers pp.4-5; Spectator 1 Aug 1931.*

Parish Church of St Comgall: 1881-99, W H Lynn: Although the Abbey church had been rebuilt and enlarged in 1832, by 1870 it had become "inconveniently crowded" and "many residing in the direction of Ballyholme made the distance their excuse for non-attendance". The Ward family became the principal donor for the new parish church of St Comgall, erected in 1881-82 to designs by W H Lynn, but the initial impulse came from the Lord Primate, Beresford, who undertook to "send, free of charge, an eminent English architect (Wm. Butterfield Esq., of London)" to advise on the feasibility of extending the old Abbey. Butterfield, who was involved in the building of St Mark's Dundela between 1876 and 1891, felt "very averse to spending money on the present church". Archbishop Beresford supported his view that a new building was desirable, suggesting casually that "you might leave the tower, etc., for the future" and so the building of St Comgall's commenced. The site had formerly been occupied by smaller buildings, including the TEMPERANCE HALL, which was also used as an infant school, and two houses that were used for a period before 1820 as the *Methodist Meeting House*.

When the church was consecrated on 8 August 1882 only the nave was complete, at the cost of £7500; the chancel, transept and spire were duly added as funds became available. Dean Maguire took the opportunity of

CASTLE STREET

climbing the scaffolding to put his hand on the top stone when the spire was completed in 1899; he "looked about him and came down quite pleased with himself". The church is built of dark basalt rubble with red sandstone dressings to windows and buttress quoins. The church is entered from a simple porch at the SW end, and the weighty tower to the NE, supported by substantial buttressing, is pierced only by small windows and the louvres of the belfry. The style is Early English Gothic with paired lancet windows along the body of the church, a rose window at the south, and some trefoil windows. Two rather impressive cylindrical stone boiler chimneys rise from the south elevation.

The interior is equally sombre, with a wide nave separated by tall sandstone columns from the aisles; the north aisle is taller and has some fine stained glass windows to F R Lepper (d.1908 - with rich vegetation in one portion and seagulls in another); Emeline Thompson (d.1897, with willowy golden haired virgins); Foster Connor; Archibald Earl of Ava (mortally wounded at Ladysmith but here represented in the shining armour more favoured in heavenly circles); and the triple lancet window in the choir to Very Rev Edward Maguire DD, who was rector of Bangor Parish from 1876 to 1903 and had overseen the building of the church. The choir has stone arcading and wooden choir stalls. The church's bells are said to be amongst the hundred heaviest in the world, the largest weighing over a ton. Steps and a ramp were added in 1998 to provide disabled access.

See *Crosbie p.24; Eakin; Ewart p.39; Handbook pp.39-40; IB 10 Oct 1901 p.888, 9 Sep 1905 p.634; Lawrence 2856, 3873, 3881, NS4406, 9542; Maguire pp.51-53, 75-76; Seyers pp.5, 21-22; Spectator 2 Oct 1986, 14 Nov 1996; WAG 394A.*

Nos.27-67: Castlemount Terrace: *c.*1895: Good long terrace of two-storey two bay stucco houses gently stepping up to follow the original level of the road which has now been cut away in front. Delicate continuous moulding over doors and windows, rendered chimney-stacks. All the original four-panel doors and horizontally-divided sash windows have now gone. The adjacent nos.3-21 were originally similar but have been even more heavily altered, with nos.23 and 25 being demolished for road widening.

No.73: Bangor Hospital: 1909-10, by Young & Mackenzie, with later additions: The original hospital is a homely red-brick Edwardian building, occupying the site of a former sand pit at the edge of the town, and looking down over what is now Ward Park. The central two-storey block, with half-hipped roof and weather-vane, has a shallow Arts and Crafts bow window below a Venetian window with a red sandstone panel decorated with ornamental swags, and is flanked by single-storey pavilions. It was built when Miss Connor gave £500 to replace the much smaller cottage hospital (see 82-86 Hamilton Road), and Lord and Lady Clanmorris gave the two-acre site rent-free. A granite plaque records the setting of the foundation stone by Miss Connor on 9 December 1909, and the hospital opened the following November. J & R Thompson were the contractors. In 1914, with

c.20 Castle Street: a terrace of urban vernacular cottages, single-storey with attics, with small-pane windows and sheeted doors below the high wallhead. Photographed about 1963, shortly before its demolition. *(H A Patton)*.

30-34 Chippendale Avenue: a surprising survival in a modern suburban context, Fern Cottage appears on the 1833 OS map when it was a small farmholding, and the central portion retains broad sash boxes. *(Peter O. Marlow)*.

a severe outbreak of scarlet fever affecting the population, a new infectious disease hospital was planned for Church Street (soldiers training at Clandeboye in 1915 for war in the trenches were confined to camp for a time by "spotted fever"), but it seems not to have proceeded. A second wing (1925, also by Young & Mackenzie) was however added later well in character, but a third wing in rustic brick and concrete (by Samuel McIlveen as nurses' homes, later adapted to outpatients accommodation) jars considerably. A decision to close the last in-patients facility was taken in 1996, but it has since expanded as a community hospital.
See *App 3119; IB, 24 Aug 1935; Lawrence 11213, 11214; PRONI D.2194; Spectator 11 Dec 1908, 7 Oct 1910, 2 Jan 1914, 30 Jan 1914, 12 Mar 1915, 1 Oct 1955, 13 June 1996.*

Castle Cottages: *c.*1880: Terrace of four stucco houses, all now altered, numbered as if coming into Bangor. Paired gables to the road, originally with label mouldings to gable windows. First known as *Ward Cottages*, they stood at the head of the wooded walk from the Castle through what is now Ward Park to the old Cottage Hospital.

Nos.38-42: This was the location of the former FIRE STATION, adapted by Stanley Devon at a cost of £3000 from what had been the gas-cleaning station.
See *IB 11 Jun 1949 p.548; Spectator 23 Dec 1950.*

Nurses' Homes: 1956, by Stanley Devon: Three stylish two-storey tenement blocks in rustic brick with brightly-painted porthole doors and hipped tiled roofs, named after local benefactors Tughan House, Hadow House and McMillan House.
See *App 3119; IB 30 Jun 1956; Spectator 26 May 1956.*

15 **Bangor High School:** 1929-31, by Castor J Love of Co Down Education Committee: Long symmetrical two-storey block with central and end pavilions set forward, in rustic brick with Westmoreland slate roof. Originally known as *Castle Street Public Elementary School*, it was designed to take 1200 pupils and be the largest elementary school in Ireland; it cost £43,000 to build, and the contractor was Thomas McKee.
See *BNL 1 Aug 1931; Hogg 25-28; N Whig 26 Aug 1931; Spectator 14 Sep 1929.*

GATE LODGE: See *Castle Park Road.*

CATHERINE PLACE: See *Dufferin Avenue.*

CEDAR MOUNT
Short road off Bryansburn Road, developed *c.*1935 on the site of a former nursery.

CENTRAL AVENUE
Straight narrow road dipping down from Main Street and up again to the top of Dufferin Avenue. It is shown on the 1833 map, and was known as *Middle Road* or *Diocesan Road* in the mid 19th century, when it led directly to

Crawfordsburn; it was only developed with buildings after 1860.
See *BHS I p.61; Seyers p.2.*

No.35: Central Gospel Hall: 1917 with front altered *c*.1995: Little roughcast gabled hall with battered shoulders, originally with small projecting porch and known as *Central Hall*.

Nos.41-45: 1997, by Hall Black Douglas for BIH Housing Association: Three storey smooth-rendered apartment block with end bays set forward. This replaced the CENTRAL BUILDINGS, which was built in 1922 as a garage for Jacob O'Neill's *Pioneer Bus* business, and had a coarse Art Deco facade appropriate to that purpose. However it also encompassed an auction mart, a school of dancing and a bakery shop. Shortly after the bus service was sold to the Road Transport Board in 1936, the upper part of the building was used by the young ladies of the Central School, whose own building was being used by the army during the war, while the lower part was used as an air raid precaution store. After the war it was converted to become the much-loved *Little Theatre*, base of the Bangor Operatic Society and capable of holding nearly 350 people in an intimate space. All the stalwarts of the Ulster stage played there, including James Young, James Ellis and Colin Blakely. Like all good theatres, it was supposed to have a ghost, and a mind reader called Zareda would hold seances to will it to appear, although the noises heard were probably just the sound of the lights cooling down. It closed in 1992 and was demolished in November 1995.
See *Spectator 30 July 1992, 9 Nov 1995.*

Nos.47-49: St John Ambulance Youth Centre: *c*.1997: Smooth-rendered shallow-gabled hall with channelled ground floor and central vehicle entrance.

Nos.51-53: *c*.1910: Pair of two-and-a-half storey roughcast houses with an oversailing porch roof like a debutante's skirt being lifted for a curtsey.

No.55: *c*.1910: Two-storey stucco double fronted house with decorative surrounds and keystones to openings. Rosemary-tiled with terracotta finials, but otherwise similar to no.64 and no.1 Primrose Avenue.

No.57: *c*.1910: Broad two-storey double fronted house with canted ground floor bays flanking doorcase; all opes segmental-headed.

Bangor Congregational Church: 1925: Built by the Bangor Christian Workers' Society as *The King's Hall*. Gabled hall with an array of first floor lancet windows and a tiny window in apex of gable.

No.64: Halliday Cottage: *c*.1910: Two-storey double-fronted stucco house with steep decorative gables and a cherubic face beaming from the keystone over the door. Similar to no.55 and to its neighbour in Primrose Avenue, but less ornate. Windows unfortunately altered, and walls covered some years ago with a heavily textured paint that obscures details.

Nos.74-78: John Gray & Co: 1986: White roughcast single-storey building with narrow windows; on dark grey brick plinth, with cupola on roof. Well

planted round, though the creation of a car park in front has necessitated demolition and the loss of the attractively dense street corner formerly here.
See *Spectator 27 Feb 1986*.

CENTRAL STREET
Small cul-de-sac off Central Avenue lined with stepped two-storey stucco terrace houses, originally known as *Vimy Ridge*. Nos.1-15 were developed about 1920, nos.10-16 during the 1930s.
See *Spectator 6 June 1953*.

CHIPPENDALE AVENUE
S-shaped road from Donaghadee Road to the East Circular Road, in existence from the early 19th century, when it was known as *Pope's Lane*, but with development mostly post-war.
See *BHS II pp.49-51*.

41 **Nos.30-34: Fern Cottage:** *c.*1800: A surprising survival in completely suburban surroundings, this occupied the same bend in the road in 1833. No.30 (Fern Cottage itself) is a two-storey roughcast house; no.32 is single-storey with a tarred tin roof (presumably once thatched), and horizontally divided sashes in broad window surrounds; no.34 has been modernised.

CHURCH HILL: See *Newtownards Road*.

CHURCH QUARTER: See *Abbey Street* and *Church Street*.

CHURCH STREET
Running from the bottom of the Newtownards Road to the Clandeboye Road, this is one of the oldest streets in the town. It was sometimes regarded as part of the Church Quarter (see *Abbey Street*), though now mostly mid- or late-Victorian terrace houses; the curving street frontage is pleasant and many of the houses could look well if consistently maintained. In the 19th century, when it consisted of single-storey weavers' cottages it was known as the *Fourth Row* (which may have been derived from Fort Row as there may have been an earth fort around the abbey, the edge of which would have come to this point) or *Four Raw*. One solitary two-storey house was known as the POOR HOUSE - it appears to have become a private residence, but may have been the Poor House recorded in the OS Memoirs as having been supported by Col Ward in 1834.

Most of the Victorian residents were weavers, some having as many as four looms, though some had no back doors, and Charlie Seyers remembered the cows having to be driven through his uncle James' kitchen, with the manure being wheeled out into the street each spring. In 1866 it was recorded that "Thomas Campbell keeps his Ass in the house". Here also at that time lived the remarkable Sammy Reavey, recalled by Charlie Seyers as wearing no boots, but a tall hat (or two) filled with coal, beef, bones, bread and carrots;

he would carry any pennies given him in his last two fingers and his water can in his first two and would sometimes stand looking out to the sea and walking backwards.
See *Eakin; Hogg 16; Minute Books; OS Mems p.24; Seyers pp.15-17, 20; Wilson p.2.*

Nos.1-17: *c.*1900: Two-storey two bay terraced stucco houses with bold string course and moulded surrounds to ground floor openings; duple sashes in segmental headed openings on first floor still present at no.9. Modern shops at nos.1 and 3.
See *Hogg 16; Welch 33.*

Nos.33-35: *c.*1800: Single-storey two bay houses, originally with Regency sashes. Now with altered doors and windows, and with raised ridge to no.33.

Nos.37-41: *c.*1890: Two-storey two bay stucco terrace houses with horizontally-divided sashes at nos.37 and 39; no.39 has four panel door, rectangular fanlight; slate roof with blue clay ridge.

Nos.47-57: *c.*1890: Similar to nos.37-41, but with wider frontages. Built by Capt Inkerman Brown.
See *BHS II p.27.*

No.91: Ebenezer Gospel Hall: 1932: Gabled roughcast hall four bays deep, with additional porch and side hall.

No.16: The Jennie Hanna Memorial Hall: *c.*1920, by Ernest L Woods: Gabled rendered hall seven bays deep with modern porch. Miss Hanna established a temperance social club for men at *The Pavilion* in Southwell Road which moved to an old cottage on this site. Her recipe of "a good fire, singing, sometimes a cup of tea, Paris buns, a currant loaf" proved so successful that the men built this hall themselves, Miss Hanna having persuaded Mr Woods to design it free of charge and then found money for the bricks.
See *Spectator 2 Dec 1958.*

Nos.52-60: Low two-storey stucco terrace at nos.52-54 with continuous hood moulding over first floor windows under tall wallhead; horizontally-divided sash windows at no.52. Then at nos.56-60 a similar terrace with continuous eyebrow moulding at ground floor.

CIRCULAR ROAD
Designed by H A Patton & Partners *c.*1970 as the last major piece of town planning before local government planners took over. Cutting through the Clandeboye Road, Newtownards Road, Bloomfield Road, Gransha Road and Donaghadee Road, it was intended to form a barrier to all further expansion of the town, but development has now hopped over and is rampant beyond it.

CLANDEBOYE
Clandeboye: 1801-04 by Robert Woodgate for James Blackwood, 2nd Baron Dufferin, with many later additions: Clandeboye is a comparatively modest

country house, originally called *Ballyleidy House*, with an unusual L-shaped plan giving it two principal elevations side by side rather than the symmetrical approach of early Georgian times. A simple portico and almost self-effacing pediment ornament the southern facade, while a shallow bow window provides the focus of the eastern one; both look out onto magnificent landscape vistas. This late Georgian simplicity probably only survived the loving attentions of its most celebrated owner, Frederick Hamilton-Temple-Blackwood, 5th Baron of Dufferin and Ava (1826-1902), because he was so busy abroad that he spent comparatively little time at his "beloved Clandeboye", as he had renamed Ballyleidy. (He was Governor-General of Canada in 1872-79 and Viceroy of India 1884-88 as well as being ambassador to Russia, Constantinople, Italy and France, being elevated to Earl and then Marquess in recognition of his achievements).

Dufferin was only 15 when he inherited the family title and estate in 1841. He commissioned William Burn in 1848-49 to come up with plans for Elizabethanising the entrance front of Clandeboye (as the Wards were doing at the same time at Bangor Castle); in the 1850s Benjamin Ferrey was asked to produce drawings for romanticising the old house along French chateau lines; and finally W H Lynn came up with elaborate proposals for baronialising it. A fascinating collection of drawings survives at Clandeboye, and it is a shame that none of the fairytale castles was ever built - although equally, a relief that the Regency Clandeboye survives. Dufferin did however leave his mark very firmly on the house, making numerous small alterations and additions. (The ever-brisk Baddeley & Ward state that "the house was modernized at the beginning of the present century, and is destitute of any architectural beauty"). Many changes were small-scale, like the often humorous titles on the mock book spines on the hidden doors concealing routes out of the library, but in the baronial passage-way which he created from the former kitchens he also designed one of the most extraordinary entrances into a grand house. The visitor passes clusters of arms and armour (pretty much up to date as it includes a World War II gas mask!), a pair of stuffed bears, Egyptian hieroglyphics, Canadian curling stones and Burmese bronze bells - climbs a narrow staircase at one end, turns sharp right and finds himself amongst the formal rooms of the house, themselves a virtual museum of artefacts collected by the well-travelled Marquess. At the foot of the staircase, narwhal tusks jut out of the stair curls, and overhead is suspended a large well-weathered log of timber "found at Spitzbergen", while the attention strays between a Burmese rain gong and a machine that promises "A Perpetual Regulation of Time". The fireplaces in the saloon are from the original 1805 house, but the leather spines on the two false doors in the Library are very much in the Marquess' taste, including one volume titled *Open Sesame*, and another the *Humour of Mr Gladstone*.

According to Harold Nicolson, Lord Dufferin's "passion for glass roofing was... uncontrolled", and he went on to adapt more of the back quarters into

Clandeboye House: Robert Woodgate's modest Georgian country house of about 1800 gives away nothing of its eclectic Victorian extensions to the rear and the exotic collections of the first Marquess within. (Hogg Collection).

The George, Clandeboye: dating from about 1858, this was Lord Dufferin's third attempt at designing a picturesque schoolhouse. It is now a restaurant, so it is surrounded by tarmac rather than grassy playgrounds. (Peter O. Marlow).

top-lit billiards rooms and cloakrooms. From the outbuildings he made a banqueting hall and a chapel, similarly adorned with items his magpie eye had identified on his travels - "a stone from Nubia, a bit of cornice from a Coptic church found in Egypt, and the shaft of an old Irish Cross he rescued from serving duty as a door stop in a kitchen garden".

The late Lord Dufferin and Lady Dufferin have done much to bring the house back from the neglect it suffered from military occupation during both world wars, restoring the buildings, acquiring suitable furniture and carpets to set off the house and its extraordinary collections. They have also added modern items, particularly paintings by Duncan Grant, Lady Dufferin's teacher, and works by David Hockney, who did a celebrated etching of the library window during a visit to the house.

There are a number of outbuildings arranged informally close to the house, many occupied as houses. Of particular interest is the main stableyard, from part of which Lord Dufferin created a **banqueting hall** containing a *salon des refusés* of the larger and less interesting family portraits, a **chapel** in one corner of the yard, and a baronial **gas house** inserted alongside it, complete with a large Burmese bell. In the centre of the yard stands a **dovecote**.

Near the chapel are an **ice house** and some formal gardens. The demesne was laid out from the early 17th century, but much of its picturesque mixture of ornamental planting, open parkland and farming dates from the early 19th century, and Terence Reeves-Smyth considers it to be the work of the Scottish landscape gardener James Fraser. However the romantic names attached to that landscape almost all emanate from the 5th Baron.

See *Baddeley p.55; Clandeboye, passim; Crosbie p.61; Dixon p.50; Eakin; Hogg H05/31/12, 12, 25; Lawrence 9526, 9528; Monuments Record; Nicolson, passim; Spectator 26 Jan 1906; WAG 2212; Young p.188.*

87 **Helen's Tower:** 1848-50, by William Burn: On a hill-top about a mile from Clandeboye towards Newtownards stands this most famous of the first Marquess' architectural achievements, a square-plan tower in basalt. Although named after his mother, Helen Sheridan, grand-daughter of the playwright Richard Brinsley Sheridan, it may not originally have been intended simply as a tribute to her. In 1847 Lord Dufferin was aware of the need to create employment to relieve the destitution of the Famine, and there may have been a social impetus behind it. Burn's perspective of 1848, in the British Architectural Library, is labelled "Gamekeeper's Tower", but in November 1850 the completed tower was christened Helen's Tower. Fitting out appears to have been a more leisurely operation, and it was not till October 1861 that it was described as "finished". Alfred Lord Tennyson responded to Lord Dufferin's request for a verse to put inside the completed tower with

> Helen's Tower here I stand,
> Dominant over sea and land.
> Son's love built me, and I hold
> Mother's love in lettered gold.

Helen was still alive at that time but more poets were asked to produce poems for Lord Dufferin, Browning and others contributing lines after her death in 1867. The Marquess was obviously very close to his mother, from whom he derived his artistic inclinations, and the Tower is a celebration rather than a funereal tribute.

Inside there are indeed a couple of rooms where a gamekeeper might live, but the room on the second floor has a heavily coffered ceiling emblazoned with heraldic devices, and the third floor is richly panelled with timber Gothic tracery including rib vaulting and barley sugar columns. There is a further small turret room on the parapets, from which a stunning view of the surrounding countryside can be seen, many of its features romantically named by the Marquess - Gauntlet Wood, Carpenter's Clump, Miss Linley's Hill, Lady Hermione's Hill, Spurs and Roses Clump, the Burma Clump, Quebec Wood, Lord Warden's Wood, Rapier Wood, Admiral's Island - and even Lord Frederick's Wood.

In 1915 and 1916, the Ulster Division was camped at Clandeboye where they trained within sight of Helen's Tower, so that it was one of the last buildings from their homeland that many of the soldiers killed in the First World War were to see. Ten thousand men from the Division went over the top at the Somme on 1 July 1916: over 5,000 were casualties, over half of them killed. For that reason it was an inspired idea to repeat the design for the monument at Thiepval, dedicated in November 1921.

See *Baddeley pp.56-57; Clandeboye pp.28-33; Crosbie p.60; Eakin; Hogg H05/31/ 13; Lawrence 4751, 4752, 12967; Spectator 1 May 1997; WAG 1458A.*

Clandeboye Lodge Hotel: 1992-94, by Alan Cook Architects for Country Inns (Ulster) Ltd: Wide gabled front with open-apex "porte cochère" in front of a lancet doorway. Attractive red brick incorporating black brick lozenges and string courses, and even a big chimney-stack, set in well-wooded car park with old trees retained.
See *UA Oct 1994 pp.20-21.*

The George: *c.*1858: Now a restaurant, this was originally built as the *National School* serving the children of the Clandeboye estate workers. It is a gabled L-plan red brick building with black lozenges and alternate black bricks in window dressings; the roof forms a catslide over the timber-columned side porch; lattice windows set in lancet recesses; now greatly extended on two sides. Lord Dufferin had commissioned earlier designs from William Burn in 1850 and Benjamin Ferrey in 1854, but this design is different again, with a touch of the estate carpenter in the rather crude arrangement of the lattice windows.
See *BHS I pp.17-18.*

Red Row and **Bangor Lodge:** see *Belfast Road.*

Other buildings associated with the estate include the eccentric Helen's Bay railway station and bridge, which are outside the scope of the present

publication. The private road to that station leaves the western side of Clandeboye via the **Demesne Bridge** over the road. Further down that road is the South or **Newtownards Lodge**, a plain rendered (but once with half-timbered upper floors) building of $c.1890$ with ribbed chimney-stack on front roof slope, and Gothic gate pillars. Dean describes other lodges of $c.1830$, 1845 and two of 1855 which are within the estate. The 1994 Blackwood Golf Centre by O'Donnell & Tuomey is in the Clandeboye estate, but outside the area under consideration in this book.
See *Crosbie p.63; Dean p.69; Eakin.*

CLANDEBOYE PLACE
Development of two-storey semis, all now modernised, laid out around a bleak pentagonal courtyard off Clandeboye Road. Developed by 1928.

CLANDEBOYE ROAD
Continuation of Church Street south-westwards across the West Circular Road to Rathgael Road. Few buildings of interest, starting with two-storey interwar terraces that give way to warehouses and recent housing estates. Charlie Seyers recalled the rural Clandeboye Road of the late 19th century, when a man called Hogg came and started a pig farm, and the two brothers Dunn clambered into their cart to do messages, wearing "Tammy Shanters... about as old as themselves". Two rises on the road were known respectively as *Widder Bray* and *Jack M'Dowell Hill*.
See *Seyers pp.42-44.*

Clandeboye Road Primary School: 1956-57, by WDR & RT Taggart: Cluster of rustic brick buildings set behind neighbouring houses and well screened from the road. Informal plan with glazed hall to one side of the entrance and larger admin building to the other.
See *Spectator 24 Aug 1957.*

St Andrew's Presbyterian Church: 1956-57, by J Tomlinson: Steeply roofed church of rustic brick laid in Flemish bond, set on hill above the road. The front gable is bisected by a vertical strip of glazing rising to a belfry that lacks a bell. Formed in this location in 1948, the congregation initially worshipped in a large Nissen hut they had obtained from the military quarters at Clandeboye.
See *Spectator 13 Apr 1957, 23 Jul 1998.*

W P Nicholson Memorial Free Presbyterian Church and Christian School: $c.1985$: Utilitarian to a fault, broad two-storey roughcast building with plastic windows and concrete tile roof.

No.244: Ava Farm: $c.1870$: Single-storey house with high wallhead, and door with fanlight in segmental-headed ope. The original farmhouse and outbuildings remain, but the name has now been adopted by an adjacent housing estate ($c.1995$) which occupies part of the former fields.
See *Seyers pp 42-44.*

Clifton Mount: See 121-125 Victoria Road.

CLIFTON ROAD

An attractive road linking the top of High Street to Ward Avenue to form an almost circular "inner ring" route round the headland between Bangor Bay and Ballyholme Bay; laid out about 1850 to serve the Ward Villas and extended as far as the Royal Ulster Yacht Club by 1900. Good stone boundary walls exist in front of many of the houses.

Nos.3-13: The Anchorage: *c.*1910, built by Capt David Lindsay: Terrace of two-storey stucco houses with paired first floor windows over canted ground floor bays.
See *BHS II pp.25-29.*

No.17: Lorne Villa: *c.*1910: Detached two-storey double-fronted stucco house with paired plain sashes over hipped ground floor bays; continuous eyebrow moulding to first floor windows.

Nos.19-21: *c.*1890: Two-storey pair of semi-detached stucco houses with unusual roundheaded niches above front doors; two-storey canted bays with horizontally-divided sash windows.

No.23: *c.*1910: Two-storey stucco detached house; segmental-headed sash windows and four-panel door; two-storey canted bays.

Nos.29-31: *c.*1890: Three-storey pair of stucco semi-detached houses with scallop-topped two-storey canted bays, and main doors at side porches; some margin-paned sashes; lean-to conservatories at first floor. This and the next pair of houses were built as the *Washington Villas.*

Nos.33-39: *c.*1900: Two pairs of substantial brick two-storey semis with broad mutual gables to front, first floor roughcast; canted bays at ground floor; first floor windows originally bipartite.

No.43: *c.*1890: Two-storey painted brick house, roughly cubical in shape but with irregular bays to front and side. Cream brick dressings to opes, which are generally segmental-headed, with horizontally-divided sash windows; tall chimneys, and weathercock (askew) over door turret. This was acquired as a meeting place by the *Church of Christ Scientist* in 1953.
See *Spectator 8 May 1959.*

No.45: *c.*1915: Picturesque red-tiled house with scalloped tile-hung gables, roof sweeping down to form verandah; lunette and porthole windows; the original windows with leaded lights have unfortunately been replaced with plastic.

Nos.69-77: Ardbraccan Terrace: *c.*1880: Terrace of two-and-a-half storey stucco houses with thistle finials to gables, and a strange concertina-like fold at the gable to Upper Clifton, where an angled plan meets a square gable; otherwise much altered, the original horizontally-divided sash windows all gone.

CLIFTON ROAD

No.79: About 1860, David Connor built near here a double-fronted property known as THE TOWER or Tower House. It was demolished about 1935 and its grounds were developed as Braemar Park.
See *Welch 36*.

53 **Nos.97-99: Ardmara:** *c*.1860: Pair of substantial two-storey semi-villas with shallow Regency-glazed bow windows looking towards the sea; in red brick with stucco dentils under eaves, a fringe of dormer windows, moulded window surrounds and quoins; large brick porches and blind windows to the road. No.97 has unfortunately been rendered. At the time of writing, this fine building, one of the earliest in Bangor and still comprising two good family homes, is sadly under threat of demolition. [Demolished February 1999.]
See *Eakin; Lawrence 3876; Seyers p.27; Spectator 10 and 17 Dec 1998, 7 and 14 Jan 1999*.

iii, 53 **No.101: Royal Ulster Yacht Club:** 1897-9, by Vincent Craig: Picturesque two- and three-storey red brick clubhouse with five-storey square tower above entrance. Red-tiled roof well equipped with finials and chimneys, half-timbered gables and dormers. An eyecatcher at the head of its peninsula set on a commanding headland, surrounded by an attractive whinstone wall, with clusters of New Zealand flax on its Seacliff Road boundary and veronica bushes at its Clifton Road side. A wonderfully varied building in what the *Irish Builder* called the "old English style of architecture", in tuck-pointed Laganvale bricks and Peake's red roofing tiles, with Arts and Crafts windows set in segmental-headed openings and a verandah overlooking the bay. A recent barrel-roofed front porch addition (by Tony Wright) unfortunately obscures the carved stone entrance. Inside, there is a very rich Edwardian interior, complete with billiard room, elaborate stair case carved with ships and flowers, the tiller from Lord Cantelupe's 84-ton yacht *Urania* (which was wrecked on the shore nearby in 1890) and splendid "Anti-Fouling and Non Contagious Closets" in the gents.

In 1866, the Marquess of Dufferin & Ava (as he was to become), recently returned from a spell as Under-Secretary of State for India, decided to revive the *Ulster Yacht Club* which had been founded in Bangor in 1806. The Marquess, who had taken his yacht *Foam* to Iceland and Spitzbergen in 1856 (as a result of which experiences he published his *Letters from High Latitudes*) was an experienced yachtsman; on his 1856 voyage he had tried a wide range of cures for seasickness including prussic acid, opium, champagne, ginger, mutton chops and tumblers of salt water. Presumably he found something effective, since his enthusiastic commodoreship of the club led to its receiving a Royal Charter in 1870, and becoming the centre for yachting near Belfast Lough. In 1889, a young architect called Vincent Craig became a member of the club, and after lengthy discussions Craig eventually became the architect for a new "Club House, Office Houses and other erections". Messrs McLaughlin & Harvey's tender for the work was accepted, and the building opened on 12 April 1899, having cost £6396 12s 1d (50% over tender price!)

Ardmara, 97-99 Clifton Road: tragically demolished in 1999, this was one of the finest early villas on Bangor's seafront, with magnificent bow windows looking across Belfast Lough. (Peter O. Marlow).

Royal Ulster Yacht Club, 101 Clifton Road: Vincent Craig's Arts and Crafts Victorian clubhouse of 1897-99 retains much of its original interior, although some of the later additions leave something to be desired. (Peter O. Marlow).

64-66 Clifton Road: like Ardmara, the Ward Villas predate the arrival of the railway in Bangor. The margin-paned windows (here unfortunately now plastic) and battlemented parapets tally with the date of 1855. (Peter O. Marlow).

Bangor Grammar School, College Avenue: the original "really handsome pile of buildings" designed by Ernest Woods in 1905 seems to have derived its inspiration from tower houses rather than the more conventional Oxbridge colleges. (Peter O. Marlow).

CLIFTON ROAD

with furnishings amounting to £1362 3s 2d.

By the end of the century the club's annual regatta attracted "all the crack boats in British waters", including those of "that boating grocer", as Kaiser Bill called Sir Thomas Lipton, who was blackballed from the Royal Yacht Squadron and therefore issued his challenge for the America's Cup from the Royal Ulster in 1898. His yacht *Shamrock I* lost, but Lipton, undaunted, issued four more challanges from the RUYC, his last being in 1929. The Club has an interesting room of Lipton memorabilia. (See also Seacourt in *Princetown Road*).

See *Eakin; Hogg 17; IB 15 Apr 1899; Lawrence 2361, 3876, 9533; Minute Books in PRONI (D.2747/1/3); Spectator 20 April 1989; Welch 36-40.*

Bangor Grammar School Prep Department: 1965, by H A Patton: Asymmetrical flat-roofed two-storey rustic brick building with eaves trim, concrete string course over ground floor. This occupies a site formerly occupied by GLENLOLA SCHOOL, including a two-storey extension by Stanley Devon in Orlit construction. Two later extensions of the Grammar School to the north have involved the demolition of nos.16-24, a group of five detached double-fronted stucco HOUSES with canted ground floor bays and label mouldings over first floor windows, built about 1900 for S McWhinney.
See *IB 31 Jul 1954; Spectator 26 Jun 1954, 6 Nov 1959, 1 Jun 1962.*

No.28: Kinnbawna: Two-storey building with original roughcast, half-hipped roof bellcast at eaves; one chimney with extraordinary top-heavy detail.

No.40: *c.1925*: Two-storey roughcast house with stucco wallhead chimneys; two-storey canted bays, one canted and one square; stained glass top-lights.

Nos.64-66: The Ward Villas: 1855, built by the Russell brothers: Pair of two-storey stucco semi-villas with gables set slightly forward carrying dated escutcheons and chimneys with tulip pots. Margin-paned windows in moulded surrounds with keystones and cornerstones; castellated bays and side porches. No.66 (known as *Eastward*) was occupied by S F Milligan, a director of Robertson Ledlie Ferguson & Co in 1910, and previously had been used by the boarders of *Dr Connolly's School*. Its present owners seem bent on removing all the mature trees that have grown up since the house was built, and there is concern at the time of writing that its demolition is being considered. The houses have been known for some time as *Eastroyd* and *Westroyd*.
See *Seyers p.27.*

54

Nos.68-70: Roslyn: *c.1880*: Two-storey three bay stucco house with central porch, curious tripartite ground floor windows and deep eaves, with giant pilasters at corners.

No.72: Dunard, previously *Tara: c.1910*: Two-storey double-fronted red brick house with trefoil-pierced crested clay ridge, red tiled roof; scalloped eaves boards. Windows partly altered, but some stained glass lights survive.

CLIFTON ROAD

Nos.74-76: Croom Villas: *c.*1895: Pair of semi-villas with broad three-storey mutual gable and central chimney-stack; partly pebbledashed.
See *Eakin; Lawrence 3876.*

Clifton Terrace: See 102-112 Seacliff Road.

Cliftonville Terrace: See 16-22 Albert Street.

COLLEGE AVENUE
Pleasant suburban street from Ballyholme Road to Shandon Drive, with hedged gardens and mostly Edwardian houses employing the then fashionable simple pebbledashed or roughcast finish that gave a more cottagey feeling than the stucco that had dominated Victorian Bangor. Being laid out in 1903, it was originally known as *College Gardens*, and derived its name from the new school that formed its focal point.

No.3: *c.*1910: Two-storey red brick house with pebbledashed first floor and rosemary-tiled roof. Upper lights to first floor windows have unusual japonaiserie margins and amethyst leaded glass upper lights to ground floor windows. Well-kept garden with unusual trees.

Nos.5-7: *c.*1910: Pair of detached two-storey stucco houses with bossed hood mouldings to first floor windows.

No.13: Bangor Grammar School: 1905 by Ernest L Woods, with later extensions: Irregular two- and three-storey building with Art Nouveau Dutch gables and castellated clock tower topped by timber belfry and weathercock (or rather shipcock), all pebbledashed with stone dressings. The right-hand portion (Crosby House, by Young & Mackenzie) which is roughcast, was built in 1914 as the headmaster's house, while to the left is the extension of 1960 (by H A Patton) which is flat-roofed, but pebbledashed to match the original building; beyond are the recent prep school and other extensions on Clifton Road. Although open to the street, the boundary is marked by a good group of trees.

The origins of the school go back to the will of the Hon Robert Ward in 1831 which left £1000 for the use of the Provost and Burgesses of Bangor "for building and endowing a school for the education of boys in Mathematics, Astronomy and Navigation, so as to qualify them as masters of foreign vessels"; this led to the building of *Bangor Endowed School* in 1856, a picturesque two-storey building with label mouldings, on the site of the present Bank of Ireland in Main Street. The school became defunct about 1893, but *Dr Connolly's Intermediate School* inherited its funding in 1900 and seems to have taken on the former school's identity. Dr Connolly's school occupied two houses in Seaforth Road (Ardmore) from 1897-1900, with boarders living at Ward Villa East on Clifton Road and their playing fields the present Kingsland Park; in 1901 they moved to nos.256-62 Seacliff Road, and in 1905 plans were approved for the present "really handsome pile of buildings" in College Avenue which were designed by Ernest L Woods, the town surveyor, and built by Hutchinson Keith, opening in September 1906. Pupil numbers grew from nearly 200 at the end of the 1930s to some 1100 by 1985.

When Sir Thomas McMullen addressed his fellow old boys in 1932, he recalled his headmaster Mr Moloney in the early days of the Bangor Endowed School as "a dear old gentleman with a white beard" who "always wrote with a quill, which spluttered all over the place". There were only twelve boys at the school then, and not much work done - "on very fine days the old doctor sent them home because it was too sunny to work, while on wet days he objected to teaching only the few boys who turned up, and sent them all home. His opinion was that school was no place for growing boys."
See *Apps 298, 697; Eakin; IB 16 Dec 1905 p.914; Lowry p.lxxxix; Milligan p.52; Morton pp.22-24; PRONI D.2194, xxxvi.3; Spectator 15 Dec 1905, 7 Sep 1906, 30 Jan 1932, 6 Nov 1959, 5 May 1961.*

No.6: *c*.1910: Simple but well-kept house with small-pane upper sashes and good stained glass at door and first floor; projecting rafters at eaves. Chimneys have vertical corbels and flat tops.

Columbia Terrace: See 59-61 Dufferin Avenue.

CONLIG
Village due south of Bangor on the old Newtownards Road and close to the former Lead Mines.
See *Old Bangor Road, Bangor Road, Main Street* and *Tower Road.*

COOTEHALL ROAD, Crawfordsburn
Rosevale Farm Cottages: *c*.1850: Two single-storey cottages with front porches.
See *Spectator 17 Sep 1998.*

COPELAND ISLANDS
The Copelands consist of three islands: Copeland Island (sometimes the *Big Island*), Lighthouse Island (formerly *Cross Island*) and *Mew Island*. According to the Ordnance Survey Memoirs a church formerly existed on the main island, presumably near the graveyard on the south of the main island and giving rise to the name of *Chapel Bay*. The islands were then owned by Mr Ker of Portavo, whose predecessor James Ross had obtained a fee farm grant of the islands and Portavo estate from the Earl of Clanbrassil, but were supposed to derive their name from a family who came to Ireland with John de Courcy. When a small coastguard vessel was moored off the island, Mr Ker "would not suffer it to remain there"; no reason was given, but it was noted that the absence of excisemen did happen to make the Copelands "a convenient receptacle for illicit goods". The Parliamentary Gazetteer of 1844 quotes an account written a century earlier describing life on the islands when the inhabitants (seventy-five in number in 1831) cultivated

harvests of "oats, barley, pease, and beans, being fertilized by an inexhaustible fund of the alga marina or sea-wreck" and kept their cattle in enclosures of sods. In 1861 the island schoolmaster kept his bed in the schoolroom and slept there, but later he lodged month about with the five families whose children he taught. About 1890 there were fifty people on the islands, with eight or nine houses apparently in splendid order. Mew Island is now a bird sanctuary, and the islands are uninhabited on any permanent basis but Milligan tells a nice story of an enthusiastic young Methodist minister paying a pastoral visit to the islands in more populous times. On landing at Chapel Bay he asked if there were any Christians on the island; to receive the reply "No - we're all Cleggs or Emersons".
See *BHS I pp.12, 20; BT 10 March 1947; Milligan p.51; OS Mems pp.21-22; Praeger pp.89-90; Spectator 7 Nov 1936, 22 Apr 1944*.

The Copelands Lighthouses: Although there had been a beacon tower on the islands from about 1714 (Harris writes of a lighthouse seventy feet high and burning one and a half tons of coal on a windy night), the first LIGHTHOUSE was built 1813-16 by the Dublin Ballast Board on Cross Island, at a cost of £9651 17s 6d, using designs by their engineer George Halpin. This produced a fixed white light that was visible for up to fifteen miles in fine weather, being supplemented by a bell in foggy conditions. The tower was abandoned when the present lighthouse was built further out on Mew Island, but Green describes keeper's dwellings, and a tower of coarse rubble masonry 26 feet square, with five foot thick walls on which the granite ashlar lighthouse would have been built. The **Mew Island Lighthouse**, a tapering tower painted black with a broad white waist-band and beacon and surrounded by a cluster of single-storey white-painted buildings, was designed by William Douglass, engineer to the Commissioners of Irish Lights, and built 1882-84. It is 123 feet high and had its own gasworks to power the foghorn and fishtail burners. In 1969 it was electrified, then in 1996 after a year as the last manned lighthouse off Northern Ireland, it was handed over to computer control.
In 1953, the *Princess Victoria* sank off the Copeland Islands with the loss of 128 lives.
See *BT 25 Aug 1993; Green pp.77-78; OS Mems p.21; Spectator 13 Oct 1928, 7 Feb 1953, 25 Aug 1994*.

CORPORATION STREET: See *Victoria Road*.

CRAWFORDSBURN
Pretty village laid out along a single main street and consisting almost entirely of well-maintained white-painted smooth-rendered houses, many with black plinth bands and glossy black doors. A pleasant variety of single- and two-storey houses survives, and the road remains narrow, houses with low curving boundary walls being at each end and consistent stone walls throughout much of the village. The roads into the village from Bangor, and particularly the

Ballyrobert Road, go through lovely wooded passages, the latter (named by Lord Dufferin *Edith of Lorne's Glen*) partly cut through rock, with ivy and ferns massing under the beech trees. Unfortunately many houses in the village now have plastic windows, and some are over-restored and twee. The celebrated Old Inn itself is dwarfed by its much bigger extensions and its car park makes a sizeable hole in the streetscape.

See *Ballymullan Road, Cootehall Road, Crawfordsburn Road* and *Main Street*.
See *Lawrence 2847, 9557*.

CRAWFORDSBURN ROAD

Continuation of the Bryansburn Road from Springhill Road past Carnalea to Crawfordsburn village, almost all open fields in 1900 and spasmodically developed through this century. In recent years there have been large developments of new houses off this road.

Presbyterian Church: See *Rathmore Road*.

Roseville Cottages: *c*.1890: Pair of one-and-a-half storey semi-detached houses, to which a third house was added followed by a later terrace of a further three. Built as labourers' cottages associated with Carnalea House.

St Galls Church: *c*.1960: Plain roughcast hall with concrete brick campanile; glazed porch set forward, with two etched glass panels.

No.173: Ballywooley House: *c*.1880: Two-storey stucco house, with horizontally divided sashes; round-headed window in gable.

No.175: Ballywooley Place: *c*.1800 and later alterations; formerly *Ballywooley House*: A long one-and-a-half storey house with narrow round-headed windows in dormers and rather grand bargeboard to gable; originally a single-storey house.

CARNALEA HOUSE: The wooded glen facing Rathmore Road indicates the position of this substantial villa which was built about 1840 looking towards the sea and with a garden to the east. About 1880 a LODGE was built at the top of the glen. In the 1950s Carnalea House was used as a temporary church for *West Church*, Bangor's sixth Presbyterian congregation, until the new church was built. Both house and lodge have now been demolished.
See *Dean p.67; Spectator 20 Oct 1961*.

Home Farm: *c*.1890: Single storey red brick building with half-timbered gables forming a formal courtyard in front of outbuildings. Square tower with weathercock and pyramidal roof.

Home Farm Lodge: *c*.1900, by Watt & Tulloch: Two-storey Arts and Crafts lodge with roughcast first floor over red brick ground floor, and half-timbered gable to road.
See *Dean p.71*.

Crawfordsburn House: 1906, by Vincent Craig: Two-storey house with additional storey lit by gabled dormers, in squared rubble sandstone with

Glendore House, Crawfordsburn Road: an unconventional stucco house with highly ambitious chimney stacks and gable doorcase, nestling in woods on the outskirts of Crawfordsburn. (Peter O. Marlow).

Burn Lodge, 200 Crawfordsburn Road: this early 19th century gate lodge to the original Crawfordsburn House has a rather Palladian appearance with its octagonal upper floor and pedimented porch. (Peter O. Marlow).

numerous chimneys, and ashlar dressings to opes and shouldered gables. Coat of arms over the entrance porch. In 1972, Charlie Munro recorded that the interior had "good cornices, doorways, staircases, decorated frieze to laylight, leaded staircase lights and vestibule screen", all in good condition. This building replaced an earlier HOUSE of *c.*1820 which stood closer to the unusual five-sided **walled garden**, and itself replaced a house of *c.*1780. There are two fine gate lodges, one to each of the earlier houses, Burn Lodge and another at the Helen's Bay entrance of the estate. The Crawfords left the house in 1934 and after the war it was donated as a hospital for TB patients, but latterly it has for many years been an old people's home; it is currently being absorbed in a new housing development. The "Crawfordsburn Fern" was discovered here, but is thought to be extinct.
See *Molloy; Spectator 13 and 20 Nov 1997; Young p.79.*

No.200: Burn Lodge: *c.*1812, by John Nash: Elegant two-storey gate lodge to the original *c.*1780 house at Crawfordsburn, with niches on sides of porch to front door, and tiny colonettes at the angles of the octagonal first floor that rises with oeil-de-boeuf windows from the main lodge.   60
See *Dean p.71.*

Windmill stump: *c.*1830: Random stone stump about thirty feet high on hillside in front of Glendore House, now roofless and almost featureless.

Glendore House: *c.*1840: Two-storey building of random rubble with lined stucco elevations and stone skews; plain double-hung sashes, some with small panes; door in gable with Ionic capitals and roll-moulded surround; bay windows to front.   60

No.212a: Glen Cottage: *c.*1820 and 1868: Two-storey roughcast house with slate roof and blue clay ridge; ground floor windows six panes over six, first floor three over three, all double-hung. Ornamental bargeboard with timber finials. Includes former coachman's house for, and contemporary with, Glen House, and stable buildings of 1868.

No.212: Glen House: *c.*1820: Large double-pile roughcast house with tall chimney-stacks, and small-pane windows, some margin-paned, set in broad surrounds.

CROFT STREET

Short curved street (the curve possibly marking part of the perimeter of earthworks around the mediaeval Abbey) from Belfast Road to Church Street. It was present in 1833, when it was actually rather longer, linking Church Street to what is now Abbey Street, but the new Belfast Road had cut through its northern end by 1858, and it went into decline (many houses being described as "in ruins") before the erection of its present buildings which date from just after 1910.

Nos.2-12: Patteson Terrace: 1912: Terrace of two-storey two bay smooth-rendered houses with extraordinary shouldered sentrybox-like door surrounds and duple first floor windows. Almost certainly named after Rev William Patteson, minister of the Second Presbyterian congregation in Bangor, who had died in 1886; he was "a familiar figure about the town in his plaid dress". See *BHS II pp.25-29; Presb Hist pp.113-15.*

CROSBY STREET
Narrow street rising from Quay Street to Holborn Avenue, much of it developed about 1890 by William Kerr. It was originally known as *Quay Place*, and then as *Crosbie Street*.
See *Hogg 44; Seyers p.12.*

Nos.17-33: *c.*1900: Terrace of two-storey stucco houses, stepped with hood mouldings over windows. A very attractive vista down the steep street to the piers and beyond to the Marine Gardens, spoilt by recent brash alterations to the lower terrace at nos.9-15.

No.10: Salvation Army Citadel: 1997, by the Boyd Partnership: Red brick building with dormers, replacing an earlier Citadel.

D

Dam Bottom: See Springfield Avenue.

DELLMOUNT AVENUE
Steep curving road off the Dellmount Road above Gransha Road, presumably developed after the war on the former lands of The Dell.

The Dell: *c.*1920: Two-storey roughcast house with two-storey canted bays linked by wrought iron balcony, with "The" and "Dell" in big capital letters on panels on each bay. Originally a farmhouse entered by a lane off the Gransha Road, the site has been occupied since before 1830.
See *BHS II pp.49-51.*

DEMESNE AVENUE
Short street off Bloomfield Road, linking to Roslyn Avenue, laid out about 1925.

No.1: *c.*1925: Tiny weatherboarded chalet with small roughcast chimneys.

No.4: Whiteoaks: *c.*1925: Pebbledashed house with half-hipped gable to street and name in plaster panel; green fish-scale slates.

DIXON AVENUE: See *Shandon Drive.*

DIXON ROAD
Cul-de-sac off Ballymacormick Road in existence shortly after 1900, and

now linking to further recent development. In 1925 it was one of the four Bangor streets to boast a house called (after Bruce Bairnsfather's famous First World War joke) *Better 'Ole*.

DONAGHADEE ROAD
Long road from the top of High Street eastwards. The OS Memoirs noted it in 1835 as being a good road with an average breadth of 35 feet, and it was the main route from Bangor to the mail port of Donaghadee. The road rises over a hump around nos.55-65, presumably formerly covering a bridge to span a stream that runs from there to the east of Fourth Avenue, and is then culverted to Folly Bridge at Ballyholme. As development proceeded further down the road in the Roaring Twenties, one couple named their house *Night Owls*.
See *OS Mems p.23*.

Nos.5-7: *c*.1935: Two-storey roughcast houses with mutual brick chimney with ornamental tall pots. Doors at sides; slate-hung gables with overhanging eaves. Roof of Westmoreland slates with blue saddleback ridges.

No.21: Fairview: *c*.1890: Two-storey double-fronted three bay stucco house at top of Fairview Gardens, with gable to road. Little cast iron gate and pillars set in privet hedge.

No.23: Lismore: *c*.1905: Two-storey double-fronted red brick house with ground-floor bay windows and red sandstone lintels. Plastic windows unfortunately detract considerably.

No.25: *c*.1895: Two-storey house with full-height bow windows. Windows and door all now plastic, and crudely re-roofed.

No.27: *c*.1895: Two-storey double-fronted stucco house with full-height canted bays. Keystones to ground floor windows, which are altered. Large extension to back.

Nos.45-47: *c*.1925: Pebbledashed house with half-hipped roof presenting a truncated gable to the road. No.45 altered, but no.47 is original with duple sash windows with strange upper sashes divided vertically into three panes.

Nos.55-65: *c*.1935: Since the road is considerably higher here than the ground to the north, these houses were developed slightly later and overcome the difference in level by means of a concrete drawbridge. The houses are pebbledashed, with plain brick basement floors below road level.

No.99: *c*.1935: Roughcast bungalow with hipped slated roof topped by a central chimney; stained glass toplights to three-light windows.

c.**No.253:** *c*.1800: Two cottages, one built of rubble stone with slate roof, the other slightly taller, mud-walled with brick wallhead; both derelict.

Nos.265-69: Aikens Cottages: *c*.1800, but altered: Single-storey cottages with good squat round gate pillars and the traditional houseleek growing on one.

DONAGHADEE ROAD

c.**No.275**: Ashleigh Cottage, formerly *Rock Cottage*: *c*.1800 but enlarged: One-and-a-half storey cottage with ornamental bargeboards to dormers, set at an angle to the road behind a high hedge.

Nos.2-22: *c*.1910: Terrace of three groups of stucco houses, mostly with two-storey canted bays and second floor windows in gables. No.2 seems to have been built as a shop, and an early shopfront survives, with elegant angle columns. Most houses altered, but no.6 is well preserved with plain sashes and four-panel door. A similar terrace stands at nos.28-34.

Nos.36-42: *c*.1935: Terrace of roughcast houses with canted bays rising to neat gables with terracotta finials. No.38 has original sash windows and stained glass door.

Savoy Hotel - See *Hamilton Road*.

No.48: Runnymede: *c*.1905: Double-fronted Edwardian house with brick ground floor and roughcast first floor. Two-storey canted bays rise to gables with heavy kneelers; escutcheons in gables; red brick gable chimneys; double-hung sashes. Local greystone wall with small gate that may have served a small older COTTAGE demolished about 1925.

Nos.54, 56: *c*.1895, for James Lenaghen: A pair of similar detached stucco double-fronted houses with full-height canted bays; four-panel doors in segmental-headed opes with fanlight and sidelights; double-hung sash windows. Before the area grew up, Mr Lenaghan could have looked from the windows of no.56 straight down the new Ward Avenue to the sea at Lukes Point.

No.58: Brookhill: *c*.1922: Roughcast bungalow, with overhanging eaves to hipped slate roof and louvred shutters to moulded window surrounds; central small-pane dormer window. Beautiful garden with mature copper beeches, and hipped garage down a hedged lane.

Nos.66-80: *c*.1905, for Nathaniel McCready: A number of two-storey double-fronted houses with ground floor bays below paired first floor windows, all unfortunately now altered in various ways. A monkey puzzle tree survives at no.70, but few of the houses retain double-hung sashes or original panelled doors, and most have been re-rendered with loss of detail. It is likely that they were all stucco, and had a local stone boundary wall such as that which survives at nos.78-80. McCready owned this land in 1893, and although he applied to develop the six houses in 1902, he only developed the core houses nos.70-76 around 1905, and no.66 was not built till about 1920.
See *App 97*.

No.92: *c*.1930: Two-storey double-fronted pebbledashed house with bold flat-topped two-storey bows; mullioned and transomed windows with curved glass to bows.

No.106: *c*.1930: Two-storey roughcast house with steep hipped roof and central chimney; two triple sashes with Arts and Crafts windows at first floor.

No.134: Bayview Cottage: Double-fronted stucco cottage with pair of finialed dormers, and canted bays at ground floor. Stucco boundary wall and pillars.

No.304: *c*.1910: Hipped-roof workers' cottage, recently modernised.

DONAGHADEE ROAD, Groomsport

Continuation of Main Street from The Lodge to join the lower Donaghadee Road from Bangor. At that junction there is a rubble-stone wall with a wonderfully spiky top, which originally surrounded a walled garden for the now Groomsport House Hotel.

No.1: The Lodge: *c*.1880: Two-storey building fronted in sandstone ashlar with ornate fretted bargeboards to gables and porch, mullioned windows, tall octagonal chimney-stacks; rear construction in yellow brick. Slightly spoilt by a squat window in the right-hand bay, and rather more so by the much-altered extension to the west. Apparently it was built as a rectory by John Perceval Maxwell of Finnebrogue near Downpatrick, but when he died shortly before its completion it became a dower house for his wife and children.
See *Lyttle p.41, UA Jun 1992 pp.12-14.*

67

No.2: Rocklands: *c*.1925: Two-storey rubble sandstone building with rosemary-tiled roof and gablets; broad battered chimney-stack rising up the E gable; impressive rubble-stone boundary wall with circular turrets at corners, running down to the rocks at *James Bay*. This may have adopted the name *Violet Lodge*, which originally belonged to Sir Robert Boag's residence on the other side of the road, now demolished.
See *Hogg H05/54/11-13.*

68

Parish Church: 1842 and later additions: A simple roughcast barn-plan church with pinkish sandstone dressings; square-plan two-stage tower above the porch and added chancel and transepts. The tower has a pointed doorcase below a shield bearing the date 1842, lancet windows at the first stage linked by a hood moulding, and topped by stone finials (now reduced from their original design) and a castellated parapet. The east gable has intermediate finials and a belfry-like top feature above a lancet with bossed hood, over triple windows. Good corbelled sandstone pillars at the gate, but unfortunately the entire churchyard is covered with tarmac.

67

The interior is simple and light, with an exposed hammerbeam roof structure supported on stone corbels over the nave; pointed arches at the crossing open into the small chancel and transepts. There is a stained glass window by Meyer & Co of Munich to William Perceval Maxwell (d.1875).

In 1838 the Down & Connor Church Accommodation Society was set up, and raised £32,000 to build and equip sixteen churches, of which this is one. Charles Lanyon made his services available to the Society and presumably was responsible for the charming small church which held its first service in 1842. It was built at a cost of £750, of which about half came from the

DONAGHADEE ROAD, GROOMSPORT

Society, on a site donated by John Waring Maxwell of Finnebrogue. When the chancel was added in 1909 with its marble and mosaic floor, the original flat ceiling was replaced by the present timber structure, and new pews were inserted. Equally extensive were the changes of 1932 when transepts were added to the designs of James A Hanna, in Conlig stone with dressings of "artificial Portland stone", and stained-glass emblems by his son Denis O'D Hanna.
See *Lawrence 9556; Welch W05/54/2.*

April Cottage: *c.*1935: A cobbled drive leads down to this rather large bungalow with leaded light dormers.

No.12: Groomsport House Hotel: *c.*1850, by James Sands for the Maxwell family: Formerly *Maxwell House*, situated in once-generous grounds at the east of the village overlooking *Cove Bay*, this is a two-storey cream sandstone building in Jacobethan style with shouldered gables to attic floor, mullioned windows with label mouldings and tall octagonal chimney-stacks; canted bays to seaward side with quatrefoil fretwork; the porch has Gothic openings and octagonal corner pillars. There is a good interior with encaustic tiles and ribbed vaulting in the hall, spacious stairhall with good panelled doors, and ornate cornices to main rooms. The **gate lodge**, like the main house, has shouldered gables and octagonal chimney-stacks, to which it adds a gabled porch with escutcheon over the Gothic door. The octagonal stone pillars to the gate screen have coronetted tops.

The house has much in common with the Ward's contemporary but rather grander Bangor Castle. It has even more in common with one of the sketches of "William Walker, architect" for Bangor Castle in the Public Record Office, and Walker set up office in Bangor during the decade to 1856. However Kenneth Robinson and Dixie Dean have both established James Sands, who was working for the Marquis of Downshire at Hillsborough, as the architect. He visited the site in January 1844, and freestone was shipped from Glasgow six months later, with the accounts concluding in 1848 at a total of over £6,000. In recent decades the building has declined somewhat, with the grounds given over to caravans, and more recently to bungaloid developments, with the gate lodge and outbuildings vacant or badly maintained. An extensive refurbishment with stone extensions at the front was opened as a hotel in 1994, but closed. The range of whinstone rubble outbuildings are in an even more distressing condition. [Currently (1999) the house and lodge are undergoing extensive work to provide apartments, including the unsightly exposure of the basement of the main house.]
See *Crosbie p.43; Dean p.78; PRONI D1556, D3244; Spectator 30 June 1994; WAG 3136.*

Andrews' Shore Field: *c.*1935: When the first edition of this book was published in 1984, this field of meadow grass was completely surrounded by gaily painted little chalets, ornamented with porches, verandahs, clapboard windows and garden gates, and each bore a romantic nomination: *Jalankayu*,

Parish Church, Donaghadee Road, Groomsport: the simple church of 1842, probably by Sir Charles Lanyon, before the 1932 additions by James A Hanna. (Peter O. Marlow).

The Lodge, Donaghadee Road, Groomsport: a handsome ashlar sandstone house with ornate bargeboards, dating from about 1880. Originally intended to be a rectory. (Peter O. Marlow).

Groomsport House, Donaghadee Road, Groomsport: the sandstone mansion designed by James Sands, photographed in 1998 between its closure as a hotel and its conversion to apartments. (Peter O. Marlow).

Rocklands, Donaghadee Road, Groomsport: the shore of Groomsport has been much built up with both bungalows and caravans since this photograph was taken in 1931. (Hogg Collection).

Zeladerg, MacRoom, St Ives or *Wilbet*. There are still a few genuine chalets, but sadly in recent years many have been replaced by mobile homes that are best described as caravans covered in pebbledash. None of these new generation chalets have names, all are bland and lacking in bright paintwork, and probably none are lived in on a permanent basis.

No.19: In the best seaside tradition of simple chalets gaining accretions through time, this one is now nearly hidden by a doubly-extended verandah, concrete balustrades, coloured light bulbs, a palm tree, and even a set of plastic chairs set out on the roof of the tiny garage alongside.

No.26: Weatherboarded chalet brightly painted in yellow with white trims, facing the sea.

DONARD AVENUE
From Rugby Avenue to Rugby Park, mostly laid out in the early 1930s.

DOWNSHIRE PARK
Short cul-de-sac off Downshire Road with houses in red brick and roughcast developed about 1910 on the lane leading to an earlier house.

Nos.1-4: *c.*1912, by E & J Byrne for John Holywood: Two varied pairs of two-storey red brick hipped semi-villas with roughcast first floor and triangular wallhead dormers; canted bays with vertically divided upper sashes and tripartite first floor windows over; original doors with Arts and Crafts glazing. See *App 594*.

No.5: *c.*1900: Two-storey asymmetrical house in red brick and roughcast with octagonal corner tower with metal finial; verandah and balcony.

DOWNSHIRE ROAD
A mature well-wooded street of good detached houses off Princetown Road, with a "garden line" of privet hedges unfortunately broken at a bungalow development between nos.13 and 21. A steep hill near no.20 is crowned by tall pine trees, and beyond Maxwell Road the road continues downhill to the sea front near Stricklands Glen.

No.1: Collingrove: *c.*1910: Two-storey pebbledashed house with stone bays at ground floor, moulded bargeboards with plaster half-timbering to gables, and Arts and Crafts windows.

No.3: The Beeches: *c.*1901: Two-storey double-fronted stucco house with canted ground floor bays. Moulded surrounds to segmental-headed opes with keystones. One of the early occupants was Thomas Wilson, later knighted, who became the first mayor of Bangor.
See *App 46; Wilson p.63.*

Nos.5-11: 1905-10: Decent detached two-storey double-fronted stucco houses with bay windows and plain double-hung sash windows complete; French doors on to small balconies over the front doors. Nos.7 and 9, and possibly

their later neighbours, were designed by Ernest L Woods in 1904.
See *Apps 190, 411.*

No.13: *c.*1930: Two-storey detached red brick house with yellow brick dressings to opes and quoins; irregular plan. Red clay ridge with trefoil piercing, polychrome brick chimney. Varied windows with stained glass upper lights, many segmental-headed with terracotta keystones. In 1921 this site and that of the adjoining bungalows was a *Lawn Tennis Ground.*

Nos.21: Highfield: *c.*1905: Irregular red brick house with roughcast upper floors, roof with half-hipped elements and bizarrely top-heavy corbelled chimney-stacks; stained glass upper lights to many windows.
See *Hogg 106-108.*

Nos.23-25: *c.*1900: Pair of irregular smooth-rendered houses with two-storey bays, on the crest of the hill. Upper floor windows mostly segmental-headed in chamfered opes, many margin-paned with tinted glass.

No.27: *c.*1905: Substantial two-storey red brick house with half-timbered gables squashed onto one another; double-hung sashes, corbelled brick chimneys. Set back from the road.

Homes of Rest: *c.*1898-1908, by W J W Roome: Three detached houses at the bottom of Downshire Road facing the sea; of two and three storeys in height, and varied designs including hipped roofs and bonnet gables, built over a period of years. These were financed by a philanthropist called Vance, although the third house was known as the *Mrs Forster Green Home for Mothers and Children.* Another home was built in Brompton Road (qv).
See *App 315; Eakin; IB 15 Sep 1900 p.482; Hogg105; Lawrence 2861; Spectator 1 and 29 Dec 1905; Wilson p.13; Young p.601.*

9 **No.2: Dalmeny Lodge:** *c.*1905, by Ernest L Woods for John Brown: Two-storey double-fronted stucco villa, with overhanging eaves; left-hand bay rectangular with fretted bargeboard, right-hand bay canted and rising to tall turret-like parapet, originally battlemented; central mahogany porch. Mature garden with palm tree, clematis and sundial. The original owner ran the two commuter boats to Belfast at the turn of the century.
See *App 202; Wilson p.62.*

Nos.4-4a Duniris and Dunraven: *c.*1889 and *c.*1900, possibly by Young and Mackenzie: Duniris was built as a *Manse* for Rev Alexander Patton in 1889-90; it is asymmetrically double-fronted, with the left-hand bay canted and roofed, and the right rectangular and rising to a gable. Dunraven was added about 1900 as the *Misses Pattons' school*, and later converted to a house.
See *NDH 22 Feb 1889.*

No.4c: Dunedin: 1961-62, by H A Patton for himself: A discreetly sited rustic brick bungalow with shallow pitched felted roof; tall gable chimney, picture windows. Mature garden with beech trees, and a pond which started life as a limepit for the building of Duniris.

Nos.8-10: *c.*1890: Pair of semidetached two-and-a-half storey stucco houses with two-storey canted bays and attic gables, set back from the road.

No.18: *c.*1890: One and a half-storey smooth-rendered villa, with cockscomb ridge, and various gables including a dormer with ornamental bargeboard.

No.20 West View: *c.*1880: Substantial two-storey stucco house with associated outbuildings, now somewhat altered, but still with rubble-stone boundary wall and outbuildings. Built by a man named Johnston who bathed at Pickie every day and "could float like a barrel".
See *Seyers p.36.*

No.28 Rathverde: *c.*1900: Two- and three-storey stucco villa at rear of steep site overlooking the sea through a fringe of rather elderly pine trees. Gable to right of entrance jettied out; windows in surrounds with cube insertions.

No.30 Thalassa: *c.*1895, for Joshua V Eves: Three-storey stucco villa with boldly projecting dentilled string courses and partly balustraded bays, situated high on a promontory between Bangor Bay and Smelt Mill Bay. It replaced an old farmhouse known as *Fulton's*. The Eves were Plymouth Brethren, but the black sheep of the family was the engineer and inventor Fred Eves, who used to go to the pub every Saturday morning in the 1950s with his colleague Marcus Patton (the author's grandfather), and when he returned the capacious pockets of his greatcoat would be filled with bottles of whiskey to keep him going for the week, so he had to move very carefully to avoid clinking them as he went up the stairs.
See *Lawrence 9532; PRONI, D.1898/1/3; Seyers p.14.*

Dudley Terrace: See 30-42 Hamilton Road.

DUFFERIN AVENUE
The section of road falling steeply from Main Street to Southwell Road existed before 1833. In the 19th century there was a stream at the bottom, beyond which there were two granite posts and an iron gate leading into fields farmed by Captain McCullough of Rathgael. Until very recently this first section, which was originally known as *Catherine Place*, was developed only on the eastern side; the section from Southwell Road to the junction with Princetown Road was almost entirely built between 1890 and 1900. The present name comes from the 1st Marquess of Dufferin and Ava. The street was developed in terraces, with few buildings of individual interest but of very consistent character; unfortunately the last twenty years has seen a large number of houses heavily and insensitively modernised.
See *Lawrence 3874, 11222, C6015.*

Dufferin Court: *c.*1985: Terrace of four rendered two-storey retail units stepping down from the railway station.

Social Security Office: *c.*1950: Rustic brick building with rosemary-tiled hipped roof behind parapet; six-pane sashes in cement moulded surrounds.

Nos.1-19: Inkerman Terrace: *c*.1890: Mostly stucco three-storey terrace houses with two-storey canted bays, many gableted or with dormers. Nos.1-3 originally brick but grotesquely rendered in the 1980s. Named after the Crimean battle of Inkerman of 1854 - but not directly: they were built many years later by Captain Inkerman Brown who was born in 1857 and named after an uncle who had been killed at Inkerman. These may have been the "first six houses in Dufferin Avenue" by Robert Neill for Robert Yarr, for which Charlie Seyers carted sand off Bangor shore at eightpence per load.
See *BHS II p.27; Lawrence 3874; Seyers p.35.*

Nos.21-31: *c*.1895: Two-and-a-half storey brick terrace of paired two bay houses; two-storey canted bays below wallhead dormers carrying a sunrise motif. Houses sit above raised gardens.
See *Lawrence 3874.*

Nos.35-49: Astoria Terrace: *c*.1880: Two-and-a-half storey terrace of stucco houses with canted bays, horizontally-divided sashes and ornamental bargeboards and finials; supported by the bookend of **no.33** which is a three-storey double-fronted house. The terrace was listed in the 1920s as *Victoria Gardens*.
See *BHS II p.27; Lawrence 3874.*

Nos.51-61: Columbia Terrace: *c*.1880: Two-and-a-half storey terrace of stucco houses with canted bays, which had ball finials to the wallhead dormers.
See *BHS II p.27; Lawrence 3874.*

Nos.63-71: Castleton Terrace: *c*.1892: Three-storey painted stucco terrace houses with two-storey canted bays; no.63, *Castleton House*, has a broader bow window, nos.65-71 have roundels above the bay windows. All originally had stout iron railings on stucco dwarf boundary walls.
See *BHS II p.27; Lawrence C6015.*

Nos.73-79: Gosford Terrace: *c*.1892: Three-storey painted stucco terrace houses with two-storey canted bays, with five-panel doors under heavily corbelled entablatures. Originally had stout iron railings on stucco dwarf boundary walls.
See *BHS II p.27; Lawrence C6015, 11222.*

Beyond no.79 was the two-story stucco PRIMROSE COTTAGE built about 1880 with ornamental bargeboards to gables and dormers, and eyebrow mouldings over windows, which stood on a block of its own between Dufferin Avenue and Princetown Road. It was demolished for road widening about 1935, and its site is now occupied by a roundabout. Its name plaque however was salvaged and applied to no.39 Ballyholme Esplanade.
See *information from Ian Wilson; Lawrence 11222.*

Nos.2-32: Catherine Place: *c*.1840: These former houses are now almost all commercialised and no longer "the swell street in Bangor" recalled by Charlie Seyers from the 1860s, neatly stuccoed with broad pedimented stone surrounds to windows and doors, and with steps up to each house from the

Catherine Place, Dufferin Avenue: photographed in its swell days around 1890, when the pavement had ornamental cobbling and several houses sported window-boxes, it is now entirely commercial. (Lawrence Collection).

Dufferin Avenue: the upper part of Dufferin Avenue photographed about 1900, when it was newly constructed. Note the chunky railings to front gardens and uniform sash windows. (Lawrence Collection).

DUFFERIN AVENUE

kidney-stone pavement. However the small-paned sash windows do remain at first-floor level at no.6, and no.14 has the ground floor almost intact. Alex Miller, who lived here in the 1860s, was a carpenter and owner of Bangor's only hearse: it had a black rocket in each corner, and a central rocket which would indicate the deceased's marital status - white if single, black if married. Until 1900 the town *Dispensary* was situated here (below Dr Russell's house in what is now the Ava Hotel), treating mainly scrofula which "medical men attribute... to the proximity to, and intercourse with, Scotland". The Dispensary had been discontinued as early as 1844 due to "unsatisfactoriness of management" and lack of funds, but presumably re-opened under Dr Russell.
See *Lawrence 3874, C2355; Hogg22; OS Mems pp.23, 25; Parl Gaz; Seyers p.2.*

Nos.36-54: Landerville Crescent: Three-storey stepped terrace curving into Southwell Road, with two-storey alternating canted and bow bays, the latter with colonettes between the windows, and the roofline broken by triangular wallhead dormers; substantial rendered chimney-stacks. David Aumonier recalls lodging at no.54 (now roughcast and in flats) when it was painted pink and the front garden was full of garden gnomes; when you walked through the door "you would have thought you were in the Highlands of Scotland" with heads of stags all over the place, but disconcertingly "a typical breakfast might have been chicken and ham pie and banana".
See *Spectator 4 Sep 1997.*

Nos.56-64: Fernville Terrace: *c*.1895: Three-storey stucco terrace with two-storey canted bays, and strapwork ornament over fluted doorcases.
See *Wilson p.88.*

73 **Nos.68-76: Belvoir Terrace:** *c*.1895: Three-storey stucco terrace houses with two-storey canted bay windows. Nos.70 and 72 are still nearly original, with plain sashes and four-panel doors. **No.66** is a later infill designed to match; **nos.78-84** probably date from the 1920s.
See *Wilson p.88.*

Nos.86-96: mostly by J Thompson, 1900: Three-storey stucco terraced houses with two-storey parapet roofed bay windows. Some still have chamfered opes, and four-panel doors in colonnetted openings with semicircular stained glass fanlights; with red and black quarry tile paths up to them, but the curly iron railings that originally ornamented the dwarf walls were removed in the last war.
See *App 22; Lawrence C6015.*

No.96a was formerly a CORNER SHOP on the flatiron site between Central Avenue and Primrose Street, demolished *c*.1985.

Nos.98-108: West End Terrace: *c*.1890: Terrace of two-and-a-half storey stucco houses with small Dutch gablets to second floor windows, stucco mouldings including escutcheons in the gablets, gable bargeboard and chequered quarry tile paths; with some unfortunate recent alterations.
See *Lawrence 5477, C6015, 11222; Wilson p.61.*

DUFFERIN TERRACE

Private road off Groomsport Road leading to the rear of Dufferin Villas, which face the seashore.

Dufferin Villas Cottage: *c.*1880: Now considerably altered and extended, this must originally have been a simple three-bay gate lodge facing the laneway, as Dufferin Cottage at the other end of the access road to Dufferin Villas still does. Some ornamental verges and vertically-divided sashes survive.

Hamilton Villa: *c.*1880: At the western end of Dufferin Villas is the slightly larger, originally stucco-fronted, gabled Hamilton Villa - now roughcast and with most windows altered. In the 1890s the Milligan family had its summer residence here: one of the daughters, Alice, co-founded the republican magazine *Shan Van Vocht*, and with her sister edited Bunting's Irish folk songs; her younger brother Charles was a local historian and mayor of North Down in the 1960s.
See *Milligan pp.1-2, 55-56*.

Dufferin Villas nos.1-8: *c.*1875-80: Row of two-storey semi-villas built in rubble stone with red-brick dressings and quoins, and stucco two-storey bay windows. The hipped roofs are fringed with dormer windows and topped by hefty chimneys, and the houses are set on imposing sites overlooking the bay. Sadly many of these fine buildings have been pebbledashed and had windows altered in recent years. The houses were the brainchild of Rev Isaac Mack, minister of First Presbyterian Church, Groomsport, who intended the income from them to benefit a Church Trust. They were incomplete when he died in 1877, but are still numbered from the Groomsport end.
See *BT 12 Mar 1998; Milligan pp.1-2, 55-56; Nelson p.25*.

Dufferin Terrace: *c.*1880: Terrace of three roughcast two-storey houses on the main lane to Dufferin Villas; with first floor windows in square dormers.

Dufferin Cottage: *c.*1875: Lined stucco one-and-a-half storey gate lodge with chamfered opes to windows on either side of pitched roof porch. Slate roof with ornamental projecting rafters. Extended at rear.
See *Dean p.74*.

Dunedin Terrace: See 6-28 Beatrice Road.

E

Edenville: See 114-120 Seacliff Road.

Elizaville: See 29-41 Brunswick Road.

ELMWOOD DRIVE
Street from Oakwood Avenue to Belfast Road, developed during the 1930s. Mostly terraced houses with running ground floor bays, interspersed with a few small bungalows of the same date.

Elsinore Terrace: See 14-20 Southwell Road.

Epworth Terrace: See 12-28 Hamilton Road.

ESPLANADE: See *Marine Esplanade* and *Quay Street.*

Eureka Terrace: See 9-21 Prospect Road.

F

FAIRFIELD ROAD
Road curving off to the east of the Gransha Road before rejoining it at Ballymagee. Formerly known as *Williamson's Lane* from the owner of the nearby farm, it was little developed before about 1925.
See *BHS II pp.49-51; Wilson p.90.*

FAIRVIEW GARDENS
Cul-de-sac off Donaghadee Road developed about 1955 on the site of former glasshouses, taking its name from *Fairview*, the house at the top of the road. Mainly two-storey semis in rustic brick with hipped rosemary-tiled roofs over canted ground floor bays, probably originally having steel windows and stained glass doors as survive at no.1.

FARNHAM PARK
A cul-de-sac developed off Farnham Road about 1900-1910. Largely Edwardian double-fronted stucco houses, mostly with two-storey canted bays, making a pleasant street, although with few individual designs.

No.7: *c.*1905: The best-preserved of a group of two-storey double-fronted stucco houses with two-storey canted bays, from nos.3-11; plain double-hung sash windows in chamfered surrounds, four-panel door with hood moulding over, dentilled cornice. No.9 is also reasonably preserved.

No.13: *c.*1905: Two-storey double-fronted stucco house with segmental-headed opes.

No.19: Park House: *c.*1910: Two-storey double-fronted stucco house with two-storey canted bays, dentilled cornice. All windows segmental-headed sashes, ground floor ones in moulded surrounds with keystones. John McMeekan, who was elected Chairman of Bangor Urban District Council

no less than twenty-two times, lived here towards the end of his life.
See *Wilson p.68*.

Nos.2-4: *c*.1905: This pair of semi-villas were formerly among the more individual designs of the street, stucco with aedicule roofs over inner bays and substantial doorcases, but re-rendering and plastic windows have wiped out much of their character.

No.20: Dun Alastair: *c*.1910: Two-storey double-fronted stucco house with two-storey hipped canted bays; door in segmental-headed moulded surround.

FARNHAM ROAD

A good varied road from Bryansburn Road to Downshire Road, of mostly Edwardian houses, many of the standard double-fronted stucco design, but several of individual interest. When it was first laid out the paving of roads was the responsibility of house-owners, and in 1904 the Spectator reported that the road was in a "disgraceful state", remarking that in wet weather it was a shame "to see the ladies' dresses trailing in the mud."
See *Lawrence 9545, 9546; Seyers p.27; Spectator 24 Jun 1904*.

No.1: Lancedean: *c*.1903, by Vincent Craig for H C Craig: Simple asymmetrical villa with tripartite first floor windows, now unfortunately replaced in plastic. In 1980 the privet hedge formed a charming arch over the gateway.
See *App 95*.

Nos.11-13: Farnham Villas: 1902, by J Fraser for R Yaw: Three-storey stucco semis with ornamental bargeboards to gable and two-storey bow windows with parapet roofs. Windows segmental-headed plain sashes, many in moulded surrounds with keystones and chamfered opes.
See *App 90*.

Nos.19-21: *c*.1905: Pair of two-storey unpainted stucco semi-villas with canted outer bays and segmental-headed doorcase in moulded surround with keystone; very shallow ground floor bays with stucco fishscale slates over. First floor windows Arts and Crafts sashes.

No.25: Arbutus: *c*.1905: Two-storey stucco house with corner turret enclosing an iron verandah, spoilt by recent plastic windows.
See *Lawrence 9546*.

No.27: Kilbrogan: *c*.1905: Two-storey double-fronted unpainted stucco house with plain sash windows and terracotta finials on canted bays.
See *Lawrence 9546*.

No.29: Pickwick House: 1902, by J C McCandless for James Campbell: Two-storey double-fronted stucco house with two-storey canted bays rising to hipped roof with finial.
See *App 87; Lawrence 9546*.

No.31: Seaview: *c*.1845: Two-storey five bay stucco house with hipped roof, decorated with stucco escutcheons above the ground-floor windows, label

mouldings and giant pilasters. Single-storey castellated wings on each side have blind-arched niches. This house has the reputation of being one of the oldest in Bangor, and indeed there was a house on the site in 1833, but it was extensively remodelled or rebuilt about 1845 into its present form.
See *Seyers p.14.*

Nos.6-8: *c.*1912, for William Bell: Pair of two-storey semi-detached houses with rosemary tiles, scalloped tile-hung gables, leaded glass, and doors with inverted semicircular lights in good Arts and Crafts manner. **No.12**, *Holly Lodge*, is in similar style but somewhat altered.
See *App 517.*

No.10: Farnham: *c.*1895: Two-storey double-fronted stucco house with two-storey canted bays; crested ridge; segmental-headed windows in moulded surrounds.

Fernville Terrace: See 56-64 Dufferin Avenue.

FIFTH AVENUE
The main road of the Bay Lands development by Gordon O'Neill meanders from Ward Avenue past Third and First Avenues down a steep hill to Second Avenue. The numbering of the Bay Lands streets seems to follow no particular order, but it has been suggested that this main street was named after Fifth Avenue in New York. Building probably got under way in 1921 and most of the houses were built before 1928.

No.1: *c.*1928: Two-storey pebbledashed house with large central gable set forward. Grey Westmoreland slates to main roof and projecting timber-mullioned ground floor bow window to gable. Good copper beech in garden.

No.11: *c.*1985: Modern infill with prominent windowless pebbledashed garage set somewhat in front of and dominating its house, and indeed (unfortunately) this part of Fifth Avenue. Not even any trees or shrubs to soften the effect.

No.13: *c.*1925: Two-storey double-fronted roughcast house with half-hipped roof and roughcast gable chimneys. First floor windows are small-pane three-light casements in corbelled-out projecting dormers. Set snugly behind a privet hedge.

No.15: Avalon: *c.*1935: Two-storey smooth-rendered house with grey-green slated hipped roof, brown saddleback ridges, and Art Deco leaded lights.

No.21: *c.*1923: Asymmetrical bungalow with roughcast walls and bellcast rosemary-tiled roof and bold battered chimneys with bulgy red pots. Laurel hedge, palm trees, cylindrical roughcast gate pillars. The former side garden, once an immaculately rolled croquet lawn, has now been developed.

No.23: Farragh: *c.*1925: Two-storey roughcast house with half-hipped rosemary-tiled roof and overhanging boarded eaves. Stained glass upper lights to mullioned windows.

31 Farnham Road: the core of this handsome house dates from before 1833, but was probably extensively remodelled and extended about 1845 when the crenellations and window labels would have been added. (Peter O. Marlow).

16 Fifth Avenue: at the focus of the Bay Lands development stands Gordon O'Neill's own house with its deep eaves, bold lack of symmetry and bizarre door porch. (Peter O. Marlow).

6 Fifth Avenue: although Bay Lands was virtually complete before the war, this single adventurous post-war addition by Henry Lynch Robinson takes it boldly into the Modern era. (Peter O. Marlow).

FIFTH AVENUE

No.25: *c.*1922, by Gordon O'Neill for Arthur Frame: Roughcast bungalow with steep, almost pyramidal, hipped roof in red fish-scale slates, with tall brick chimney alongside a tiny canted dormer over the central castellated ground floor bay. Plastic windows.

No.6: 1951, by Henry Lynch Robinson for Mrs McMillen and Miss Forsythe: 80 Modernist two-storey smooth-rendered house with monopitch roof projecting jauntily at the front like a peaked cap, and three full-height vertical features - a rustic brick chimney, bow-fronted reeded glass staircase turret, and a set of vertical rails on the north side. The only immediately post-war building in Baylands, but well fitted on its site. Now sadly with plastic windows, and the privet hedge opened up at the side to give car access.
See *App 5586; BT 6 June 1996.*

No.14: *c.*1923: Two-storey roughcast house with steep hipped roof and central roughcast chimney. Running bay along front; porch with timber column stuck at corner.

No.16: Windrum House: *c.*1921, by Gordon O'Neill for himself: Rather 79 heavy asymmetrical smooth rendered house with walls pierced by corbelled slightly projecting windows, a porthole window and a barrel-roofed porch over a nicely detailed leaded-light door and sidelights. Important fir tree at the bottom of the sloping lawn. O'Neill was the architect for the Bay Lands development, and his own house was one of the largest, as well as being placed centrally. It is distinctly more Modernist and less folksy than many of O'Neill's other designs for Bay Lands, and reminiscent of his design for the Spectator Buildings at 109 Main Street.
See *App 1028.*

No.18: The White House: *c.*1921, by Gordon O'Neill: O'Neill's "design no.16" has a half-hipped roof and a balcony at first-floor level. Block-bonded quoins to angles and windows, which have very small-paned top-lights.
See *App 1017.*

No.28: *c.*1925-26, by Gordon O'Neill for James Shields: Asymmetrical bungalow with pantile roof in a mixture of half-hips, gables and verandahs; complete with portholes, bellcast roof and cast iron gatepillars.
See *App 2113.*

FIRST AVENUE

Short road from Ballyholme Road to Fifth Avenue, part of the *Bay Lands* development, with Gordon O'Neill's own house in Fifth Avenue looking from its raised site down the axis of the Avenue towards the sea. Apart from the recent bungalows, development was complete by 1930.

No.9: Tavistock: *c.*1921, probably by Gordon O'Neill: Two-storey hipped 83 roughcast house with L-shaped plan at the junction of First and Fifth Avenues, so that the chamfered gabled corner between the two gables containing the front door addresses both streets, and the timber gate in the privet hedge is at

the junction. A small but mature garden, somewhat broken up by a car opening and extension along Fifth Avenue.
See *Spectator Directory 1922 and 1923.*

Nos.2-4: See-Sea House: *c.*1940: Utilitarian and box-like three-storey red-brick building with exposed concrete lintels. No.2 is a shop with plastic windows above, no.4 still has timber three-light windows with stained glass upper lights, along with a castellated brick garage.

Folly Bridge: See Ballyholme Esplanade.

FRONT STREET: Probably an early name for *Queen's Parade.*

FISHER'S HILL: See *Victoria Road.*

FOURTH AVENUE

One of the main streets in *Bay Lands*, rising steeply from Third Avenue and dropping down with a sharp turn to join Second Avenue. Development started about 1922, and continued well into the 1930s.

Nos.37-39: *c.*1935: Two-storey roughcast semis with red slated roof and mutual brick chimney. Oriel windows with stained glass toplights and sashes with small-pane toplights. A similar but less intact pair stands at **nos.27-29**.

No.16: *c.*1935: Red brick bungalow with half-hipped red roof with bellcast over front. Gable set forward with duple window; Arts and Crafts windows; tall chimneys on front slopes of gables. Vast palm trees.

No.18: *c.*1935: Two-storey pebbledashed house with quoins and trims; gable with terracotta finial rising from canted bay. Stained glass door with fanlight and sidelights.

No.28: *c.*1935: Roughcast bungalow with red slated roof forming front verandah, and terracotta finial to central half-timbered blind gable. Canted bay oriels at ground floor supported on consoles; two tall chimneys. Good copper beech. Currently under threat of redevelopment.
See *Spectator 23 Apr 1998.*

Nos.40-42: *c.*1925: Handsome pair of one-and-a-half and two-storey roughcast semis with red tiled roof, porches and oriels and Venetian first-floor windows.

No.44: *c.*1925: Irregular bungalow with half-hipped pantile roof; now with plastic windows.

No.48: *c.*1935: Roughcast bungalow with hipped red roof and low roughcast chimneys. Placed at an angle to street (as is no.17 Second Avenue) to face the bottom of Fifth Avenue. Monkey puzzle in front garden. A recent extension to the north has been squeezed on to a tight site with considerable discretion, using roughcast, pantile roof and dormer windows to keep the scale down.

9 First Avenue: despite recently altered windows, this design of about 1921 still makes a strong Janus-like statement at the corner of First and Fifth Avenues, with the recessed doorcase looking to the garden gate. (Peter O. Marlow).

40-42 Fourth Avenue: roughcast semis with Venetian windows and catslide roofs, typical of the Bay Lands cottage style but less altered than most. The variety of houses in Bay Lands is in strong contrast with modern developments. (Peter O. Marlow).

G

GLENBURN PARK
L-shaped road from Ashley Drive to Ashley Park, laid out about 1935.
Ballyholme Primary School: 1953, by WDR & RT Taggart: Rustic brick building with steel windows and concrete tile roof; square tower over central entrance.
See *Spectator 25 July 1953*.

GODFREY AVENUE
Street from Ballyholme Esplanade to Groomsport Road, mostly developed about 1930.
No.2: *c.*1905: One-and-a-half storey stucco villa with central verandah carried round to form a canted bay window on one corner and corbelled on heavy timberwork and columns; deep eaves with exposed rafters, deep-set dormer windows. Smartly painted with hood mouldings picked out, but sadly most of the windows now changed to plastic.

Golf Terrace: See 88-94 Hamilton Road.

Gosford Terrace: See 73-79 Dufferin Avenue.

GRANSHA ROAD
Continuation of Castle Street south-east out of Bangor to Six Road Ends, by-passed beyond the Circular Road and known as the *Old Gransha Road*. The road was in existence before 1830. A channelled stream runs into Ward Park alongside the road near the tennis courts, in which despite its rather stagnant appearance small fish still dart. Development on the road is mostly post 1930 and undistinguished.
See *Seyers p.33*.

Granville Terrace: See 76-82 Princetown Road.

GRAY'S HILL
A steep street rising from Queen's Parade to Princetown Road; in existence in 1833, but the present houses are mostly late 19th century in date. The name supposedly came from the original developer of the street, whose houses were taken down and rebuilt by his son.
See *Lawrence 3879, 11633, 12246; Seyers pp.35-36; Welch 3, 5, 7, 26*.

No.1: 1890: Three-storey stucco building with corner oriel turret. In Edwardian times, this was the studio of the photographer William Abernethy.
See *Lawrence 3879, 12775; Welch 7*.

Nos.3-5: *c.*1890: Roughcast (formerly stucco) three-storey building, with canted two-storey bays and two Dutch dormers with semicircular-headed windows.
See *Lawrence 3879, 12775; Welch 7.*

No.7: *c.*1890: Three-storey two bay stucco building with keystones to moulded surrounds; originally Regency-glazed.
See *Lawrence 3879, 12246, 12775; Welch 7.*

Nos.9-11: Auberge, Grainger's: *c.*1890: This three-storey building was originally of cream brick with red brick dressings, and unusually it has basements and steps up to panelled front doors. Unfortunately rendered and altered *c.*1990, and further altered *c.*1995 with fully glazed ground floors to bay windows. About 1900, when it was known as *Bella Vista*, it became the premises of Miss Murphy's *Ladies' Collegiate School*, later taken over by the Misses Weir, who moved to Pickie in 1919.
See *Lawrence 3879, 11633, 12246, 12775; Spectator 3 May 1963; Welch 7.*

Nos.19-23: Ainsley Engraving, Collectables: Three-storey six bay terrace with very shallow rectangular oriels with twinned basket-headed windows. Ground floor now completely tile-clad.

Nos.27-29: Pollock's, Barber's Workshop: *c.*1885: Three-storey stucco building with segmental-headed sashes in moulded opes to upper stories; reasonable shopfront with slim columns and aediculed fascia-stops.

Nos.31-33: Picture House: *c.*1900: Three-storey six bay building with heavy corbelled fascia above ground floor; possibly built as a guest house.

No.35: *c.*1870: Two-storey two bay stucco house.

Nos.45-47: *c.*1900: Two-and-a-half storey stucco houses with dormers.

Nos.57-59: Goldenage: *c.*1910: Two-and-a-half storey stucco houses; rather delicate dormers with acroteria and glazed sides.

Nos.4-12: Winifred Terrace: *c.*1908: Terrace of two-and-a-half storey stucco houses.

No.16: *c.*1900: Antiques: Three-storey two bay stucco building with good shopfront.

No.18: Classics: *c.*1890: Three-storey three bay stucco building with semicircular pediment breaking the roofline and fine bulky corbels to shopfront.

No.58: *c.*1900, probably by Nicholas Fitzsimons: Irregular red brick house neatly occupying a corner site on a steep hill, with a variety of gables and dormers, tall corbelled brick chimneys above stucco ground floor, with rosemary-tiled porch tucked into an internal corner.
See *Lawrence 5477; Perspective Jan 1996 p.60.*

GREENMOUNT AVENUE

Cul-de-sac off Manse Road (in the former grounds of the manse), developed about 1935 with pebbledashed houses. **Greenmount Court**, a building of *c*.1978 by Peter Davidson which stands at the entrance, was originally red-brick, but has been rendered.

GROOMSPORT

Groomsport is thought to be essentially a Viking settlement, with the foundations of the present pier probably going back to that time. Its name was originally *Graham's port*, *Gregory's port* or *Gilgroomes port*. The Raven map shows it in 1625 to have been a village of some 120 inhabitants occupying widely-spaced cottages along both sides of the present Main Street with a spur inland; there was a large house on the site of the present Groomsport House, and the harbour with its pier and sheltering shoal and promontory contained a number of vessels: the road from Bangor to Donaghadee followed the coastline and there was a rabbit warren to the South.

The village's chief claim to fame is that the Duke of Schomberg reputedly landed here on 13 August 1689 with an army of 10,000 men and that he wintered in the vicinity before being joined by William at Carrickfergus in June the following year on their way to the Boyne. (There is evidence, however, to suggest that his landing place may have been nearer to Bangor.) The Parliamentary Gazetteer in 1844 recorded that the harbour, 1000 feet long and 500 wide was "very safe, though shallow", being protected by a rocky shoal called *Cockle Island*, and that some eighty men operated nine half-decked vessels and eight open sailboats from it. In 1831 the population was 408; by 1898, the number had fallen to 284, and Praeger noted a village street occupied by "whitewashed cottages, a few lodging houses, a Presbyterian church, and a couple of public-houses", with a "picturesque little harbour, filled with brown-sailed trawlers and open fishing-boats", and until a couple of decades ago it had changed very little. Unfortunately during the 1960s there was a spate of demolitions of the small cottages that still remained, and the main street is now open to the sea; the pier is no longer approached by a lane but overlooked by a rather bleak field of grass and concrete brick with a few isolated and weather-beaten trees, redeemed by the surviving cottages at Cockle Row. The bay itself, with a healthy stock of yachts and dinghies, remains very picturesque.

See *Bangor Road*, *Donaghadee Road*, *Harbour Road*, *Main Street*, *The Hill*, *The Point*, *Springwell Drive*, *Springwell Road* and *Copeland Islands*.
See *Hogg H05/54/10, 11; Lawrence 2365, 4736, C5052; Praeger p.91; Welch.*

GROOMSPORT ROAD

Long rather rambling road branching off the Donaghadee Road and running to the Groomsport roundabout; laid out about 1840 but little developed till after 1900.

Nos.33-35: Sunbeam Cottages: *c*.1880: Low two-storey stucco houses, whose early date is suggested by the irregular fenestration; return into

Main Street: The Bangor Endowed School was built in 1856, its mullioned windows with Tudor labels paying tribute to the Oxbridge colleges. It served as the town hall for thirty years before its demolition in the 1930s. (NDHC).

105 Groomsport Road: a terrace house of about 1870, with margin-paned double-hung sash windows. (Peter O. Marlow).

Helen's Tower, Clandeboye: Lord Dufferin's tribute to his mother, later the model for the Thiepval memorial. (Hogg Collection).

GROOMSPORT ROAD

Waverley Drive even lower than main building. No.33 unfortunately altered considerably.
See *Milligan p.1*

Nos.61-63: Pair of semi-detached roughcast houses. Similar examples survive at nos.45-57, but only no.63 is still complete, with canted ground floor bay with stained glass top lights under verandah with red diamond slates.

Nos.91-99: *c.*1920: Rosemary Fegan, Winemark, -, Sheridan, Spats dry cleaners, Post Office: Two-storey row of shops with decent roughcast first floor and alternate rectangular half-timbered and canted bays; double-hung windows with smaller upper sash.
See *Hogg 6*.

Nos.101-105: Auburn Terrace: *c.*1870: Butcher, First Trust, house: Terrace of three comparatively early houses, only no.105 now unaltered, though in poor condition. Two-storey stucco with canted ground floor bay; four-panel door with fanlight in corbelled doorcase; windows double-hung and margin-paned. Roof slated with crested blue clay ridge, moulded eaves corbel above lower moulded string course; corbelled stucco gable chimneys, with octagonal pots. No.101 has the remains of a (now unusual) vitaglass and chrome shopfront, but was once the home of Florence M Wilson, authoress of the poem *The Man from God Knows Where*.
See *Milligan p.1; Spectator 9 Nov 1946*.

No.107: *c.*1910: Detached two-storey house with unpainted roughcast first floor over smooth red brick ground floor. Gabled bay set forward; porch, with stained glass toplights, set in the angle between gable and main front. Rosemary-tiled roof with exposed rafter ends.

Nos.109-111: Sheridan Villas: *c.*1910: Pair of two-storey semi-detached Arts and Crafts houses; rosemary-tiled roof with end gables set slightly forward, and cast iron spandrels to verandahs; brick and roughcast. Very bulgy pots on bulging chimney-stacks.

Ballyholme Bridge: When the road was laid out about 1840, the stream running down from the nearby mills (see *Bellevue*) had to be forded, and although the bridge is level and invisible to the passing motorist, there is a small river running under the road to a dell in the garden of no.125a.

No.143: Brooklands: *c.*1910: Large Queen Anne house, rendered with striped corners and jettied timber gable, deep hipped rosemary-tiled roof with projecting rafters, set back from the road with a generous and well-wooded garden.

No.145: Red Hall: *c.*1920: An aptly-named large red brick house, built close to the shore with a long front garden.

Dufferin Villas: See *Dufferin Terrace*.

No.165: Bayview Resource Centre: *c.*1965: Bland roughcast complex with flat roofs and aluminium windows, until recently *The Banks* old people's home, which was closed down about 1995. In 1833, BANKS COTTAGE stood here, and it presumably survived till the present building was erected on its site; a short section of flint-topped stone walling lining Banks Lane, which gave access to it, probably survives from the old cottage.

St Columbanus Church: 1939, by R Sharpe Hill of Belfast: A small church of coursed creamy-brown rubble with string courses of red sandstone on basalt plinth, with a campanile nestling alongside a semicircular apse at the NE corner. The interior is very comfortable with warm brick walls and apse painted sky blue. Foundation stone laid 6 July 1939. Church hall of 1957 by WDR & RT Taggart.
See *App 4731; Spectator 13 April 1957.*

Nos.90-104: *c.*1925: Four pairs of two-storey hipped pebbledashed semis with canted and gabled inner bays, and Moorish arches for recessed doors, double-hung sashes. Nos.98/100 and 102/104 are still virtually intact.

No.106: *c.*1930: Two-storey roughcast house with hipped slate roof and central chimney. Remarkable shell-shaped portico to front door.

Nos.122-128: Brooklyn Villas: *c.*1910: Two pairs of tall three-storey stucco semi-villas with two-storey canted bays.

No.130: A pair of two-storey double-fronted stucco houses with canted bays and segmental-headed moulded doorcase. Originally sharing a rather grand entrance, which now leads to a new estate called *Pinewood.*

No.160 Craigalea: *c.*1905: Two-storey stucco double-fronted house with balustraded balcony over dormers between asymmetrical bays; at end of long well-wooded drive.

GROVEHILL GARDENS
Narrow road off Donaghadee Road connecting to Thornleigh Gardens, consisting mostly of short two-storey pebbledashed terraces laid out around 1935.

GROVE PARK
From Donaghadee Road to Bellevue, a mixture of two-storey and bungalow development laid out between 1925 and about 1935, with distinctive variegated pebbledash often combined with string courses of plain cement rendering.

H

HAMILTON ROAD
A gently-curving road opened up in the 1890s to join the top of Main Street with the Ballyholme Road, originally called *Hamilton Street.* Until some years ago a HORSE TROUGH stood outside the church railings at the junction

HAMILTON ROAD

of Hamilton Road and Castle Street; when horses stopped patronising it, it used to be a cheerful sight filled with flowers but it has now been removed to the courtyard in the Heritage Centre.
See *Lawrence 3873, 9542.*

Good Templar Hall: 1872: Smooth-rendered gabled hall with small hood-moulded lancet windows and door, and cast iron finial at gable. Over the years the hall has served various functions: the Bangor Harmonic Society used to rehearse here in the 1900s, and it acted as a *court house* until it was requisitioned by the Ministry of Food in 1939.
See *Spectator 24 Sept 1955.*

93 **Bangor Orange Hall:** dated 1872: Smooth lined-rendered two-storey gabled porch with central lancet window sharing hood-moulding with blind niches on either side, and a pair of bright orange sheeted doors. The main hall, originally slated but now tiled, is set behind the porch and not very visible from the street. Apparently the Orangemen built the hall themselves, and a unique feature of it was the highly effective sound insulation between floors "which enabled a church or Lodge meeting to be conducted upstairs while the local Lambeg drum enthusiasts practised in the room below".
See *Spectator 11 Jan 1958, 14 Nov 1985.*

Nos.1-3: *c.*1895: Class, Marlowe: Pair of houses now combined with linked shopfront.

Wesley Centenary Methodist Church: 1891, by J P Philips: Symmetrical gabled design in uncoursed squared and rusticated dark (originally cream!) stone with red sandstone dressings. The church was enlarged in 1912, and extended to the west in 1964 by Gordon McKnight, the house that had for some years been its Manse being demolished to make way for the new Minor Hall. With twin entrance doors, and side buttresses terminating in finials. The Epworth Hall at the rear was built in 1924, and a Dunlop Room (named after Rev Dunlop who carried out his greatly appreciated pastoral visits on a tricycle) was added in 1932.

The original meeting place of Bangor Methodists was a hall called Bethel formed out of two small houses at the junction of Hamilton Road and Castle Street, which had "an unceiled roof and an earthen floor" and no fireplace. Amongst the New Connexion Methodists who met there were some vigorous characters, such as Bob Neill, a joiner who when alone used to have long conversations with the Devil "whom he upbraided in no uncertain fashion", Capt John Nicholson who objected so much to the introduction of a harmonium that he sat through each hymn with a large handkerchief tied tightly over his ears "in order to shut out the roaring of the bull", and the sexton Dick Carr who had neither the ability to read nor an ear for music, and used to bawl his hymns out heartily, with his hymn book often held upside-down. Although the Wesleyan Methodists had built the original meeting house in Queen's Parade (qv), they were ousted from it by the New Connexion Methodists, and

found "the people of Bangor were nearly as hard as the cold rocks of Bangor's coast". However during the Methodist revival of 1859, Rev Nicholson "broke the rocks in the name of the Lord", and when the centenary of Wesley's death came in 1891 they built this church, acquiring the adjoining house as a manse.
See *Haire, passim; Lawrence 9542; Spectator 12 June 1964.*

Nos.7-11: *c.*1900: Terrace of three-storey houses, originally with two-storey canted bays but now with ground floors shopped.
See *Lawrence 9542.*

Nos.13-21: Ava Terrace: *c.*1890: Irregular terrace of two-storey houses with varied dormers and bargeboards. Ground floors now converted to shops, but moulded door surrounds with keystone survive at nos.17 and 19.
See *Lawrence 9542.*

Nos.23-35: Ava Terrace *c.*1890: Terrace of two-storey houses with two-storey canted bays and round-headed second floor windows. In 1892 a Mr Tilson purchased the "desirably-situated dwelling house" at no.25, then "in an excellent state of repair", for the sum of £300. Ground floors recently changed to shops, the combination of nos.31 and 33 having a particularly clumsy frontage.
See *BNL 22 Oct 1892; Lawrence 3873.*

Evangelical Presbyterian Church: Formerly the *Borough Gymnasium.*

Earl Haig Memorial Hall: 1932-33, by L H Hodgins for Royal British Legion: Symmetrical two-storey building in red brick with rendered bays rising to flat-topped gables; central bay set forward with pitched gable. The builder, John Boyd & Sons, used ex-service men in the construction "as far as possible". Designed to be "a place of rest and recreation to those who survived the Great War", it was opened in July 1933 by Lord Bangor, who was "somewhat late" for the ceremony as he had sailed up from Strangford in his yacht.
See *Spectator 8 July 1933.*

Nos.43-69: Hamilton Hall: 1948-49, by J A Gaw: Designed as an annex to the Technical School in the Carnegie Library, the "First Aluminium School in Northern Ireland" is a corrugated aluminum shed, glazed along the sides and with a roof cantilevered out on fins placed between the window bays.
See *Spectator 21 May 1949.*

Hamilton Road Presbyterian Church: 1898-99 by W J W Roome, completed 1964-66 by Gordon McKnight: Irregular church of rusticated stone ("Scrabo stone shoddy dressings") with central glazed roof lantern. Bangor's Third Presbyterian congregation was set up in 1897, "in order to form an additional regiment for our King, the better to fight His battles" - this did not refer to the impending Boer War - and initially held services in the Good Templar Hall, laying the foundation stone here in June 1898. The church was originally built for £4000 by McLaughlin & Harvey, taking over from local builder James Colville who had apparently grown alarmed that the congregation had

HAMILTON ROAD

failed to raise the requisite cash. It opened on 10 September 1899 without the planned vestibule and tower, which were added to the south in 1966. The "very beautiful and commodious hall" at the rear was erected for Rev J Millar Craig to the designs of Ferguson & McIlveen, and is built of Laganvale brick with stone dressings and a steel frame. In opening it in 1932, Sir Thomas McMullan wondered whether the "vast storage accom-modation beneath the platform" was for a wine cellar or "a depository for uninteresting sermons" (he thought not).
See *Eakin; IB 1 Oct 1899; Lawrence 11234; Presb Hist pp.115-16; Seyers p.35; Spectator 3 Dec 1932, 17 Sept 1949; Wilson p.56.*

Nos.71-85: *c.*1900: Four pairs of two-and-a-half storey semi-detached houses of slightly varied design, with neat porches at nos.75-77 (by J J O'Shea) and nos.79-81, and bold volutes to the dormers of nos.71-73.

Nos.107-109: *c.*1910: Two-storey semi-detached stucco houses with bay-windowed gables flanking central enclosed verandah; segmental-headed windows with Arts and Crafts glazing.

Nos.123-125: Windsor Villas: *c.*1900: Pair of two-storey stucco semi-villas with full-height projecting bays to front and gable. Central attic gablet with segmental-headed plain sash windows.

St Comgall's Parish Church - see *Castle Street.*

No.2: Masonic Hall: *c.*1880: Striking ornately-decorated stucco building, two storeys in height and five bays wide. The central doorway has Corinthian pilasters supporting a segmental pediment, set back in the centre and supported on consoles; five- and six-pointed stars above it, and in the central bay at first-floor level a Bible is depicted in stucco, open at Psalm CXXXIII ("Behold how good and how pleasant it is for brethren to dwell together in unity") with a compass and dividers marking the place. The rich skyline is dominated by curly Dutch gables and the scrolled base of the central chimney (sadly removed about 1980) bearing the monogram BMH (Bangor Masonic Hall?), with corner urns to the balustraded parapet; the elevation to Ruby Street also good. The *Bangor Season* of 1885 records a "handsome new hall" in Hamilton Street which would be this structure; it was built with the £900 result of an appeal, topped up by Lord Clanmorris.
See *Lawrence 3873, 9542; Lyttle; Spectator 7 Feb 1952.*

Dufferin Memorial Hall: 1905, by Young & Mackenzie of Belfast: A complicated brick gable-fronted hall with red Dumfries sandstone dressings and low flanking copper cupolas, giving a strangely fierce and oppressive effect. The monogram DA above the door and coat-of-arms in the gable record the memorial to the Marquess of Dufferin & Ava; the Marchioness laid the foundation stone on 2 February 1905. Funds had been raised and tenders received in 1902. The contractor was local builder J H Savage, whose unconventional building methods included the setting up of a 42ft log to winch the hammerbeams into place, then walking up it "nailing footholds as

Orange Hall, Hamilton Road: a cross between a church and a fortress, with Tudor labels and Gothic tracery. This, and the Good Templar Hall to the left, date from 1872. Photograph taken in 1968. (Peter O. Marlow).

Masonic Hall, 2 Hamilton Road: the "handsome new hall" of about 1880 carries masonic emblems in its elaborate stuccowork, and is set off by curly Dutch gables. The central gable used to carry a tall chimney. The Dufferin Hall to the left dates from 1905. (Peter O. Marlow).

Carnegie Library, Hamilton Road: Ernest Woods' elegant building with shouldered gables, Arts and Crafts windows and perky lantern still serves its original purpose. (Peter O. Marlow).

82-86 Hamilton Road: the original cottage hospital was established in 1869 by Harriet Ward but replaced by the present hospital in 1910, and has since become three compact houses. (Peter O. Marlow).

he mounts". The outstanding debt on the hall was "liquidated" by a Neptune Bazaar with stalls decorated as grottos and caves, and music by the Cingalee band. Although built as a parish hall, it has served many public purposes, including duty as the *court house* during the Second World War.
See *App 224; Hogg 31; IB 6 Nov 1902 p.1472, 2 Dec 1905 p.870; Lawrence 3873; Spectator 24 Nov 1905, 2 Dec 1939, 24 Sep 1955, 16 Aug 1963; Seyers p.36.*

Nos.4-6: *c.*1900, probably by John O'Shea of Belfast: Pair of stucco semi-detached houses with chamfered openings; Dutch gables with moulded skews, and pillared stucco boundary walls. Some detail has been lost on no.6 (now *The Green Bicycle* tea room) by the application of textured paint.
See *Lawrence 3873.*

Nos.8-10: *c.*1900, probably by John O'Shea of Belfast: Pair of stucco semi-detached houses with chamfered openings. Textured paint has damaged no.8 too, but it retains the original sash windows.

Nos.12-28: Epworth Terrace: *c.*1901, by Ernest L Woods for James Newell: Three-storey stucco terrace with two-storey canted bays and dentilled cornice. *Rollo House* at no.12 acts as a slightly larger book-end for the group.
See *App 67.*

Nos.30-42: Dudley Terrace: *c.*1910: Three-storey terrace with two-storey canted bays, and consoles at doors supporting elaborate triangular pediments. Despite increasing commercialisation of the ground floors, nos.36 and 38 still have well-preserved frontages. A pair of semi-detached HOUSES of similar date, with verandahed ground floor, formerly stood on the next corner of Park Drive, but were demolished for road-widening *c.*1980. (Karl Smyth has suggested that Dudley Terrace was actually nos.14-28, but there is conflicting information).
See *Lawrence 11219.*

Ward Park: In 1909 Bangor Council acquired the site of Bryce's Brickworks, and Cheal's Nurseries created Ward Park, with its small stream widening into the lower pond beloved of small boys with yachts and the upper ponds (complete with twentieth-century crannogs) beloved of ducks, geese and toddlers with bags of crumbs for the birds. There are good groups of trees, not least the horse chestnuts at the Hamilton Road entrance. The original iron entrance gates (seen in a Lawrence photograph) disappeared during the war, and a dovecote forming part of the aviaries has gone more recently, but the park has some interesting features dotted through it. A well behind the culvert at the lower pond was known as *My Lady's Well*, and was reputed to be a magnesium well; it can no longer be seen and has presumably been buried beneath the ramped pathways under the speading chestnuts at the entrance to the park. The **bowling pavilion** dates from 1914 when the greens were set up; the octagonal **bandstand** near the pergolas is probably of similar date, and was moved there to make room for the War Memorial. There is a pleasant quaintness about much of the Park's layout, with its little bridges and miniature zoological gardens, although the Walt Disney-like Rabbit House

HAMILTON ROAD

has been rebuilt to a more functional design, and the wandering guinea-fowl seem to have flown.
See *Crosbie pp.22-23; Eakin; Lawrence 11230, 11231, 11234-7; Milligan p.30; WAG 3107-8; Wilson pp.18-19.*

War Memorial: 1925-27, by T Eyre Macklin: A willowy bronze Erin placing a palm-frond at the foot of a white stone obelisk on which the proposed bronze Victory never took up residence. The memorial cost £3000 and was unveiled in 1927.
See *Eakin; Hogg 23; Spectator 2 May 1925, 28 May 1927.*

U-boat gun: Taken from the German submarine UB19 and given to Bangor by the Admiralty in 1919 in memory of the Hon Barry Bingham, who won a VC at the Battle of Jutland when he commanded the destroyer *Nestor* - a monument much appreciated by children as a climbing frame.

No.78: Carnegie Library: 1908-10, by Ernest L Woods: A perky symmetrical brick building with Arts and Crafts windows providing a generous light to the first floor behind its Dutch gables - the upper floor was originally the Municipal Technical School - and a charming roof lantern on the main ridge; Penrhyn slates. Stone pilasters in the gables drop to form reveals and cills to the first-floor windows, while the ground floor windows conversely have stone cills and heads without reveals. There is a deep curving stone canopy to the entrance.

As early as 1837 there was a library "consisting of a limited number of religious and moral works" which edifying institution was founded by Mrs John Ward and supported by subscription, and later there were reading rooms on the site of the present Ulster Bank. The millionaire philanthropist Andrew Carnegie offered £1500 for the construction of a free library provided a free site was provided. The Hon Somerset Ward provided the site and the borough surveyor Ernest L Woods was preparing plans in 1903, with the enthusiastic backing of the Council which felt that Bangor needed "a free library to bring them up on even lines with other seaside resorts." However the Council dithered about their requirements and wanted to incorporate a technical school in the building as well, at which Carnegie took offence and reduced his offer to £1250. It was finally built, to revised plans of Ernest L Woods, by William Dowling & Sons, and the Marquess of Londonderry opened the building on 8 January 1910. Today, as well as the continuing library use, the Seacourt Print Workshop in the basement offers printmaking facilities to artists.
See *Bill of Quantities at Bangor Castle; Eakin; IB 5 Oct 1907; Lawrence 11238; OS Mems p.24; Spectator 24 Jun 1904, 20 Sept 1907, 24 Jan 1908, 14 Jan 1910.*

Nos.82-86: 1869: The former *Cottage Hospital*, established in 1869 by Harriet Ward, which was closed on the opening of the Castle Street hospital in 1910; now a group of three two-storey houses built in local brick with cement quoins and hood mouldings over the windows. Projecting porches, sheeted overhanging eaves and tall brick chimneys add to the cottage appearance.
See *Lawrence 11238.*

Nos.88-94: Golf Terrace: *c*.1905: Terrace of three-storey stucco houses, two of which are now clad in artificial stone.
See *Lawrence 11238*.

Tonic Fold: 1994-96, by Knox & Markwell for Glenbrook Homes and Fold Housing Association: Institutional-looking residential block with three-storey frontage to Hamilton Road and two-storey elsewhere, red brick over channelled rendered ground floor; shallow-pitched concrete tile roof with a plethora of vents and veluxes; plastic windows; built at a cost of £3.3m. Somewhat softened by grass verges, and young trees along the Hamilton Road. Two interesting buildings preceded the present one here:

BANGOR GOLF CLUB: 1903-04, by Ernest L Woods: The original Bangor Golf Club, a two-storeyed red Laganvale brick building with pine verandah, was the first building on this site, with a small golf course laid out on what had formerly been fields. The first floor contained billiard and smoke rooms, and there was a resident caretaker. The foundation stone was laid by Miss Connor in December 1903, and the builder was J & R Thompson of Belfast. When the club expanded and moved further out to Broadway, the building became first *Aubrey House* private school, and then *Connor House*, the preparatory department of Bangor Grammar School, until it was demolished in 1970. By that time its carefully tended swards had become a cinder playground, and were latterly part of the Tonic's car park.
See *App 144; Eakin; IB 16 Jan, 30 Jan 1904; Lawrence 9529-30; Wilson pp.18, 23*.

THE TONIC CINEMA: 1936, by John McBride Neill: Although the golf club became a school, the old course was valuable development land, and was snapped up as the site of probably the most celebrated cinema to be built in Northern Ireland. A stunning inter-war design, the Tonic was the finest example of cinema architecture in the province, and, seating 2,250 people, the second largest cinema in Ireland when it was built. The design was symmetrical, the entrance podium flanked by lower blocks containing ground-floor shops and first-floor flats; and the main auditorium rose behind, with curved staircase turrets and a small balconette outside the Rewind Room of the Operating Box to enable the projectionist to take a breather. Apart from the sides, which were undecorated, each of these elements was striped with alternating horizontal bands of white Snowcrete and rustic brick. Once past the two pay-boxes at the top of the entrance steps, the audience found themselves in an "enormous and luxurious foyer, with thick carpets, big, comfortable sofas and easy chairs". Upstairs there was a restaurant where waitresses would serve a mixed grill or coffee and cakes on a white linen cloth, and where dances would be held on Saturday nights.

A planning application in 1912 for a portico to the Picture House in Main Street for "Messrs the Irish Electric Palaces Ltd" shows that Bangor had already entered the cinema age at that time, only three years after the first cinema in Ireland opened in Dublin in 1909; by 1920 the Picture House had moved to the Windsor Bar building in Quay Street, but there was the Adelphi

HAMILTON ROAD

Kinema in Main Street and by 1935 the Queen's Cinema on Queen's Parade. Two brothers, Jacob and John O'Neill, were running the Pioneer Bus Service and an associated funeral undertakers during the twenties; John broke away from his brother to found the Tonic Bus Service (which ran from Bangor to Donaghadee, supposedly a profoundly healthy journey). When the N I Road Transport Board was set up to take over the bus services in 1935 John decided to invest the anticipated proceeds in the cinema which he reckoned Bangor needed, and commissioned a young architect called John McBride Neill to design it. Neill, practising in Belfast, was making a speciality of cinemas, and the Tonic was to be his third to open in seven months. The cinema cost £76,000, and the contractors were Sloan Bros of Belfast.

Bangor's "imposing and beautiful new cinema" was opened by Viscount Bangor "in breezy style" on 6 July 1936, and the "size, magnificence and artistic beauty" of the building impressed all those present. Harry Wingfield performed on the three-keyboard Compton organ with its illuminated rising console that magically changed colours, and the cinema opened with Ronald Coleman in *The Man Who Broke the Bank at Monte Carlo*. The walls were decorated "in the cinema manner, with sprayed textual (sic) metallic paint... with bars of bold contrasting colours... applied in a manner suggesting the futuristic, but not the futuristic that appals".

After its heyday in the thirties and forties, the Tonic continued to be a focus of local entertainment in the fifties, with children's matinées on Saturday mornings and live shows. About then a gentleman named Joe Blake used to sit in the front row "wearing women's clothes and eating fish and chips and a raw turnip", but he would not have been a typical customer! If the fire siren was sounding in Castle Street, the projectionist would flash lights on to the curtain to alert any firemen present. At the time of Queen Elizabeth's coronation in 1953, few people had televisions in their own homes, and the Tonic brought in no less than five of the rival machines so that their clientele could enjoy the snowy pictures live. In the late 1960s, the Curran Group of Companies, which had acquired the Tonic, was taken over by Rank, who modernised it and renamed it the *Odeon*; in 1974 it changed hands again and reverted to its old name, but its interior had been gutted and its days as a cinema were numbered. It closed in 1983; the famous organ was rescued and is now at Gransha Boy's High School. After some years with an uncertain future, the Tonic was gutted by a huge fire in June 1992, and demolished immediately afterwards.

See *Apps 4019, 4100; BT 29 Oct 1983; Builder May 1996 pp.113-18; NDH 24 June 1992; Spectator 4 and 11 July 1936, 21 Jan 1956, 27 Feb 1992, 25 June 1992, 2 July 1992, 6 April 1995.*

Bangor Crown Buildings: *c*.1950: Long narrow flat-roofed building set behind high mesh railings, with rooftop balcony like a high-level exercise yard. This site was formerly occupied by the BANGOR STEAM LAUNDRY, a mundane shed with factory chimney where James McMurray employed 120

Tonic Cinema, Hamilton Road: John McBride Neill's masterpiece in its maturity, photographed in 1954. The curvaceous walls with their brick trims, rising tier on tier above the shops, were demolished in 1992. (Donald Finlay).

The Savoy, Hamilton Road: while the Tonic has been replaced by sheltered housing, happily the Savoy has become sheltered housing, and Neill's other major building of the 1930s still survives. (Peter O. Marlow).

HAMILTON ROAD

women before it was destroyed in a blaze in 1909. Its successor, the Bangor Hemstitching Works, also carried out some laundry work and was burnt in a later fire in 1965.
See *Spectator 5 and 12 Feb 1909.*

Hamilton Road Baptist Church: 1922 and 1995: Two red brick gabled halls linked by a glazed porch; the right-hand gable is essentially an earlier brick and pebbledashed church of 1922 that was probably designed by Ivor Beaumont. The site was previously an open grassed area on which the Bangor Laundry laid out linen to bleach in the sunshine. Before its erection, the congregation met from 1908 in the Orange Hall on Hamilton Road.
See *Hogg 68; Spectator 10 Dec 1998.*

99 **Nos.116-120: Savoy Hotel:** 1932, by Robert N Savage; dramatically refaced and extended by John McBride Neill 1933; converted for Royal British Legion Housing Association by Robinson Patterson Partnership 1989-90: Four-storey International-Style building with strong horizontal emphasis and curved corner, smooth rendered walls alternating with bands of glazing - themselves originally divided into horizontal bands. The front elevation on Donaghadee Road has a vertical centrepiece above a flat-roofed porch topped with the name SAVOY in large lettering. The original Savoy was built by R N Savage of Bangor for Mr J Gaston of Northern Ireland Tours, who ran bus tours and kept the hotel open for the 13-week summer season. Although described at its opening as "a palatial building" it had rather a meagre vertical fenestration which gave rise to the local nickname of "Sing-Sing", and the following year Gaston decided to enlarge and remodel the hotel, using the young architect John McBride Neill of Belfast, who was shortly after to design the Tonic Cinema. Neill's bold and simple refacing gave the hotel a very stylish image which was, however, slightly let down by a view of the unchanged back and far side. Nevertheless, there was a sprung floor, and "an orchestra, composed of violin, 'cello, piano and drums". In the course of the recent conversion the windows were replaced, but unfortunately without putting back Neill's horizontal fenestration. The conversion was however generally sympathetic, and it is unfortunate that the opportunity was not taken to convert the Tonic in a similar manner. As it is, this is sheltered housing with style, in what is probably the most distinctive building from the 1930s surviving in the province.
See *Apps 3335, 4816; Boucher; Eakin; Spectator 28 May 1932, 13 May 1933; UA Jul-Aug 1990 pp.26-29.*

No.122: Fairview House: *c.*1900: Two-storey double-fronted stucco house with canted bays, surrounded by important trees.

HARBOUR ROAD, Groomsport
Short road from Main Street to the harbour.

Harbour: A modest rubble-stone structure of some antiquity, recently partly filled in to provide car parking. The pier, which is largely of rubble

construction, was damaged in high seas in 1996.
See *Crosbie pp.37, 38; Hogg H05/54/11; Lawrence 6199, 10194; Spectator 28 Aug 1959, 18 Jan 1996; WAG 1293, 2156; Welch W05/54/1.*

Lifeboat Station: 1884: A plain building in split-stone rubble with shouldered gables and quoins in dressed pinkish sandstone, now gratuitously adorned with curvaceous rendered extensions that obscure its original simplicity (not to mention its datestone). Built for Groomsport's first lifeboat (the *Florence*, presented by the Crimean Lady with the Lamp herself), and closed in 1920 when the last, the *Chapman*, was retired, it continued to provide a shelter for the boatmen who in 1925 were "often to be found standing in a group on the leeward side of the building, smoking, chatting and spinning yarns."
See *Lyttle p.40; Spectator 26 Dec 1925.*

Cockle Row: possibly 18th century: Two single-storey cottages were saved from demolition in the 1960s by Bangor Art Club. Only one is still thatched, but they are all that remain of the little fishermen's cottages that originally clustered round Groomsport harbour and along Main Street. Without them, and the nearby trees from former back gardens, the harbour area would be much the poorer, and the recent renovations to open the building regularly to the public as a branch of North Down Heritage Centre were very welcome; one cottage is said to be a forge dating back before 1631, where Schomberg's troops re-shoed their horses on landing. *Cockle Island* is an area of rocks in Groomsport Bay, from which mollusc gathering is no longer encouraged.
See *BT 17 Apr 1997; Crosbie p.37; Lawrence 2365; Spectator 31 Aug 1962, 5 July 1963, 15 May 1997; WAG 2156.*

Harryville Terrace: See 75-83 High Street.

Hatfield Terrace: See 42-52 Holborn Avenue.

HAZELBROOK AVENUE
Off Clandeboye Road. Two-storey terrace houses with exposed rafters, mostly roughcast over stucco ground floors. Developed about 1930.

HAZELDENE AVENUE
Street of mostly pebbledashed two-storey semis with projecting gables, laid out from Hazeldene Drive to Broadway about 1935. Probably the latest and least interesting of the Hazeldene streets, made worse by loss of hedges and almost all original doors and windows.

HAZELDENE DRIVE
When Bangor Golf Club moved from Hamilton Road to Broadway, formerly open land was developed with the Hazeldene and Moira streets. This one was laid out from Donaghadee Road to Hazeldene Gardens in 1932 when it was initially called *Bay Lands Seventh Avenue*, and the houses here have much in common with Baylands. Fully developed by 1939.

HAZELDENE DRIVE

Nos.2-4: *c*.1935: Two-storey rendered semis with Arts and Crafts windows; tiny mutual roof dormer and triangular porch.

HAZELDENE GARDENS
Cul-de-sac off Broadway leading to Hazeldene Drive, this first of the Hazeldene streets to be developed was mostly built around 1930.

No.5: *c*.1930: Hipped roof bungalow with terracotta crested ridge; ogival frame to recessed front door; Art Deco top lights to square bays; block-bonded quoins now painted out.

Nos.10-12, 14-16: *c*.1930: Two pairs of semi-bungalows with quoins inlaid with coloured glass mosaic (now mostly painted out).

HAZELDENE PARK
Cul-de-sac off Hazeldene Drive, developed during the 1930s. Potentially rather a charming street of small bungalows laid out on a gentle slope, it has been spoilt by indiscriminate "improvements" to doors and windows, and other than at nos.1 and 6 very little original detail survives.

Hibernia Terrace: See 14-18 Seacliff Road.

HIGH STREET
The second commercial street of the town, present on the Raven map of 1625 and known till the early part of this century as *Ballymagee Street*, it is on a steep hill with a vista across the Bay to the Marine Esplanade. The lower part of the street has been commercial for a hundred years, but in the 19th century "a great many of the people living in this street held a few acres of land and kept a cow and pigs" and the upper part was largely residential up until the last few decades. The upper floors of the stepped terraced buildings (still mostly two or three-storey in height) retain much of their style however, and the serried rows of slate roofs and chimneys have plenty of character. At the lower end, both corners (with Bridge Street and Quay Street) are beautifully handled with curved-fronted buildings of approximately equal height; however the recent automobile-inspired gash around no.30 is another matter altogether.
See *Crosbie p.25; Eakin; Lawrence 11218; Seyers pp.5-7, 9; WAG 3296.*

Nos.1-3: Atlas Taxis, Piccolo Pizzeria: *c*.1890: Three-storey two bay stucco buildings, originally houses; no.1 with high wallhead and no.3 with round-headed second floor windows.
See *Eakin; Lawrence 4721, 11218.*

Nos.5-11: The Twilight Zone, Khyber Pass Kebab House: *c*.1890: Three-storey stucco building five bays wide with central door between two shopfronts. No.5 had heavily carved swags of fruit on the shop fascia consoles until about 1990; ornamental mouldings to upper windows, including stucco aedicule roofs with fish-scale slates above first-floor windows. Opened as the *Dufferin Restaurant Hotel, Dining and Billiard Rooms,* "erected in the

HIGH STREET

most substantial and careful manner". In 1905, J Camlin & Co's White House here was selling the "Celebrated Shamrock Unshrinkable Underclothing".
See *BN 14 Oct 1892; Lawrence 11218.*

Nos.13-15: *c.*1970: The Penny Whistle and The Garage: Rendered three-storey two bay building replacing a stucco building of similar size with moulded surrounds and keystones to upper floor windows.

Nos.17-21: Café Céol: 1995, by David Wilson for Bill Wolsey: Strikingly modern design with timber ground floor screen that can be opened up like a stage set in good weather, and a giant orange-painted steel column set in a recess to support a small gable; first floor rather blank, concealing something called the Boom Boom Room.

This replaces all but fragments of two earlier buildings, a plain three-storey stucco one at nos.17-19, and at no.21 Capt Montgomery's THE ULSTER ARMS (which perhaps started life as the *Empire Hotel*), which had elegant round-headed dormers projecting from a balustraded parapet, shallow bow oriels at first floor and a central round-headed doorcase. The OLD COTTON MILL (see introduction) was built behind no.21 about 1800 by George Hannay; in the 1850s it was destroyed by fire, and was used for a while as a store; and Green recorded that some walls still remained standing in 1963. Indeed, some of the "old, incredibly thick masonry walls which partitioned the interior" and were removed during the present refurbishment may have dated from the days of the old mill.
See *Green pp.5, 10, 28-9; Perspective March 1996 pp.28-34; Seyers p.5; UA Jan 1996 pp.14-17.*

Nos.23-31: Baillie's, Optometrist: *c.*1920: Terrace of two-storey rendered shops originally with shallow bow oriel windows at first floor and dormer windows set into mansard roof.

No.35: Ormeau Arms: *c.*1890: Two-storey stucco public house with chamfered corner; pilastered surrounds to ground floor windows, and etched glass doors.
See *Spectator 9 May 1996.*

Nos.37-39: McElhill's: *c.*1880, altered *c.*1990: Recently "restored" building with sashes divided vertically where previously they were divided horizontally, and smooth render where it was previously channeled; perfectly pleasant, but if they were determined to create an olde-worlde character from scratch why didn't they pick a building that had already been wrecked?

No.41: Jenny Watts: *c.*1870, altered by Raymond Leith Partnership 1984: Two-storey roughcast pub with deep plinth. For many years this was *The Old House at Home*, by which name it was known in the 1880s, but its history goes back further than that, and possibly even to the 1780 claimed by the present owners.
See *Seyers p.9.*

Nos.43-47, 49-51: Nautical World, Bangor Guide Dogs Fund: *c.*1880: Two-

HIGH STREET

storey smooth-rendered houses, with stone-walled carriageway entrance at No.47. Original sash windows were horizontally-divided. Built by Saddler Johnston, who carted the stones for them from Conlig.
See *Seyers p.9.*

Nos.53-55: Underground Music, The Edge, Steinway: *c.*1890: Two-storey five-bay stucco building with high wallhead and segmental-headed first-floor windows with keystones. In 1885 this was the address of James Crosbie, Fine Art Designer and Embroiderer, who carried a "Choice Stock of hand-embroidered pillow cases, handkerchiefs, underclothing, etc, etc" and employed over a hundred people in the cottage industry.
See *Bassett pp.289-91; Lyttle.*

Nos.57-59: Maud's, Shop Mobility, Bangor Colour Copy Centre: *c.*1880: Two-storey shops with stucco first floor with quoins; shared carriageway.

Nos.61-67: Home Link and Taj Tandoori; Training & Employment Agency: *c.*1975: Pair of modern buildings, of unsympathetic design and material, replacing two pairs of four bay stucco houses. The previous building at nos.65-67 had panelled pilasters at either side.

No.69: Select: *c.*1965: Two-storey flat-roofed building.

No.71: Red Cross: *c.*1890: Two-storey three bay stucco building with hood mouldings over first floor windows and richly dentilled fascia and corbels above aluminium shopfront.

No.73: S & M Furniture: Considerably altered two-storey building.

Nos.75-83: Harryville Terrace: Ashleigh Photography, Shine, Vacant, ESE Wholesale, Insurance: 1902, by Henry T Fulton: Stepped terrace of three-storey stucco houses with moulded surrounds to windows, ground floors mostly altered to modern shops. The last remaining house at no.81 still had sash windows and a hood moulding over the ground floor window in 1997.
See *App 58; Crosbie p.25; WAG 3296.*

No.85-87: Motortune: *c.*1890: Two buildings combined in one bland frontage *c.*1968.

No.89: Toucan Wine: Single-storey house, formerly with a simple entablature and steps up to the door, converted to shop *c.*1990.

No.91: Volunteer Bureau, Cradles: *c.*1920: Two-storey three bay building with painted block-bonded brick surrounds to first floor windows.

No.93: King Fu: *c.*1890: Double-fronted stucco building with bargeboarded dormers, but ground floor commercialised.

Nos.95-97: Kevin Kahan: *c.*1900: Two-storey stucco building with hood-mouldings over first floor windows, modern shopfronts.

Nos.99-107: Shop Electric: *c.*1965: Two-storey box with oversized windows.

Nos.109-111: Sampson's, Feherty Travel: *c.*1890: Two-storey four bay stucco building with dentilled eaves, hood-mouldings over first floor windows, and

21 High Street: the simple but gracious facade of the Ulster Arms probably dated from about 1890, and its oriel windows were imitated in the later building on the right. It was demolished about 1990.   (H A Patton).

2 High Street: a very distinctive curved-plan building which forms the entrance up High Street, matched by a non-identical twin on the other side of the street. In 1900 this was the Stag's Head.   (Peter O. Marlow).

10-14 High Street: a splendid group of stucco buildings from the late 19th century cluster at the bottom of High Street, from the simplicity of the Fish Hall through no.10 with its central triplet window to no.6. (H A Patton).

6-8 High Street: so often Victorian buildings are at their most exuberant near the skyline, and this is no exception with its bold (and strange) date lettering, balustrades and finials, arcades of windows, and oriels with stucco fishscale slates. (Peter O. Marlow).

HIGH STREET

modern shops.

No.113: Praxis: *c*.1965: Totally unvarnished three-storey concrete brick box.

No.115: Hong Kong Palace: *c*.1890: Two-storey stucco building similar to Nos.109-111.

Nos.117-119: Crisp & Dry, Marquis of Queensberry: *c*.1910: Two-and-a-half storey stucco four bay building with (originally frilly) bargeboards to dormers and corbelled cornice over modern shops.

Nos.121-121a: Events, Bangor Lighting Centre: *c*.1960: Three-storey weatherboarded building.

Nos.123-129: Snacksville, Chemist, Shoreline: *c*.1950: Two-storey red brick terrace of shops.

No.2: Rose and Chandlers: *c*.1860: Quadrant-plan two-storey rendered building forming curved corner to Bridge Street. In 1885 the pub was called the *Stag's Head*, where could be had "first class accommodation, prompt attention, and a genuine article, at a moderate charge", and it had a handsome balustraded parapet. The pub has been through several changes of name: in 1908 it was the *Criterion*; and since the 1960s, the *Lightship Bar*, the *Seafront Inn*, and *Silvers* have all been tried.
See *Eakin; Lawrence 2367, 4727 etc; Lyttle; Spectator 26 Apr 1963.*

Nos.6-8: Storeys, Save the Children: 1891: Ornate three-storey stucco building dated in small central pediment. Balustraded parapet with urns on dentilled cornice; second floor windows in groups of three round-headed windows above the shallow first floor bow oriels which have stucco fishscale roofs; intricate ornament of scrolls and vegetables between the second floor spandrels and the cornice and central monogram (probably CN for Charles Neill) between the triplet windows; pilasters to sides. In 1908 this was the *Alexandra Hotel*. Modern shopfronts.
See *Lawrence 2367.*

Nos.10-12: Bangor House: Atlas Taxis, Pages: 1891: Three-storey stucco building dated in pediment over central triplet windows at first floor. Bracketed cornice, pilasters up each side of building; segmental-headed windows at second floor alternate with smaller blind pediments, while first floor windows are grouped in pairs flanking a triplet, with monograms (possibly JL?) over each outer pair and more elaborate consoled central feature with stucco fishscale roof; remains of original shopfronts at ground floor.
See *Lawrence 11218.*

No.14: Central Fish Hall: *c*.1890: Two-storey two bay stucco building with tiled shop that once sported a large golden fish.
See *Lawrence 11218.*

Nos.16-18: J A Johnston: *c*.1850 and later alterations: Low asymmetrical two-storey house and shop, stucco-fronted with unusual mouldings to windows and remains of corbelled shopfront behind modern fascia.
See *Lawrence 11218.*

HIGH STREET

No.20: Bamboo Tree, Bangor Fish Co: *c.*1930: Two-storey stucco building with central Venetian window at first floor, parapet front to hipped roof, aluminium shopfront.

Wolsey's: *c.*1880: Two-storey building in stucco with quoins, originally a four bay house with quoins and Ards doorway, with a fake Victorian shopfront by Raymond Leith Partnership, added about 1983.
See *Lawrence 11218.*

Argos: *c.*1990: Two-storey rendered and gabled building curving into Bingham Street. A banal design, but then this is really only the back door of the Flagship Centre in Main Street.

This development and the adjacent widening of Bingham Street involved the destruction in 1988 of three 19th century buildings and the GAS WORKS showroom (part of which, prior to the building of the Hamilton Road library, was the *Technical School*), a two-storey six bay channelled rendered building with small-paned steel windows in the first floor windows and a central flat pediment erected about 1910. Behind them towered the massive if utilitarian steel and brick structure of the gas works retorts which were cleared about 1975. The gas works committee first met in 1854, the directors including R E Ward, Robert Neill, John Brown and John McNab, with Henry McFall as Secretary. In the 1860s the gas works was a small place where two men did all the work, but the Town Commissioners took it over for the sum of £750 in 1882, and a few years later it was feeding eighty street lamps.
See *Bassett p.287; IB 6 Jun 1901 p.757; photos in NDHC; Seyers pp.5, 44.*

Nos.40-42: Davidsons, Sweet Inspiration: *c.*1890: Two two-storey shops with stucco first floor; recent neo-Gothic shopfront to no.42.

Nos.44: Donegan's: *c.*1970: A modern building, formerly covered in fake half-timbering; given a neo-Victorian frontage about 1990 for which it does not look the worse. Previously merely called the *Cartwheel*, it has now gone the whole hog and parked a complete cart on its balustraded roof.

Nos.46-48: Bank of Ireland: *c.*1970: Rustic brick building.

Nos.50-56: Graham's, Greenwood, Peppers: *c.*1890: Stepped terrace of two-storey stucco shops with some reasonable shopfronts. At no.54 for some sixty years Cecil Greenwood fulfilled his ambition of owning a shop, selling postcards, souvenirs and practical jokes in variable taste. A flashy dresser in his youth, with loud pullovers, thin moustache and a beret, he also ran a touring concert party. [Greenwood's shop is now altered].
See *Spectator 19 Mar 1998.*

Nos.58-60: XtraVision, Grate Style, John Heyes, Covent Garden: *c.*1900: Two-and-a-half storey terrace with stucco upper floors, sash windows and wallhead dormers with bargeboards. Modern shops, one advertising its wares with a fireplace stuck to the wall at first floor.

Nos.62-68: Vacant, Campbell, Colour Sound, McMurray's: *c.*1890: Stepped terrace of two-storey shops with sash windows at first floor.

Nos.70-74: Muriel Terrace: Wray Parke, Care and Share, Knightsbridge: *c.*1890: Three-storey stucco terrace (built in two stages) with moulded surrounds and stone cills to upper windows; good shopfront at no.74, and name plaque.

Nos.76-78: Dragon House, Furniture Nook: *c.*1890: Pair of two-storey stucco houses, with third storey added to no.76. The render has recently been stripped from no.78 to expose small rubble stonework with brick reveals.

Nos.80-82: Criterion Wallpapers, Pharmacy: *c.*1900: Three-storey stucco building with shops.

Nos.84-94: Morrow, Undercover, Stratford, Wing Wah, vacant: *c.*1890: Terrace of two-storey buildings with ground floor shops, all much altered.

Nos.96-98: Down Diving Services, McVeigh: *c.*1870: Two-storey stucco terrace with ground floor shops.

Nos.100-126: Irene Terrace: *c.*1900: Stepped terrace of two-storey stucco buildings, all but one now with commercial ground floors. In 1968, all but six were still domestic, with plain sashes at first floor above ground floor bow windows topped with a coronet of iron railings, which still survives at no.102. Thomas Hanna, the creator of three charming naïve paintings of Bangor in the 19th century, lived much of his life here.
See *BHS II pp.10-14; Wilson p.82.*

No.128: *c.*1950: On Your Bike: Two-and-a-half storey red brick building with long plastic shop fascia.

HOLBORN AVENUE

From High Street to Seacliff Road. At one time known as *Union Street* (possibly a name connected with the linen embroidery industry for which Bangor was famous), it is named *Holborn Street* on the 1903 OS map, and it had been given its present name by 1907. The lower portion to Seacliff Road was originally called *Well Road*. A Mr Cairns who lived here had patented a non-puncturable tyre for which he was seeking a manufacturer in 1904. The tyre was "utterly devoid of air" but rather expensive to make as it contained a hundred and fifty small springs. In 1906, "William Walshe Architect" lived at no.49.
See *Seyers p.7, Spectator 17 Jun 1904.*

Nos.9-45: Morrowvale Terrace: *c.*1900: Terrace of simple two-storey two bay red brick houses, unfortunately all but one now pebbledashed, covered in artificial stone, or painted.
See *BHS II p.27.*

Nos.47-59: Holborn Terrace: *c.*1885: Terrace of two-and-a-half storey stucco houses with wallhead dormers and two-storey canted bays with good detailing including dentilled cornice, ball finials and five-panel doors. The bay at no.47 is double width and very unusual.

c.30 High Street: houses from about 1860 photographed in 1915. Note the subtle variation in cornice and window heights, and the leviathan retorts of the gas works looming up behind. Cleared in 1988. (NDHC).

6-18 Holborn Avenue: single-storey blackstone houses, originally with brick window and door reveals now hidden in a variety of unsatisfactory ways. These must have been quite typical of mid-19th century housing in the town. (Peter O. Marlow).

Holborn Hall: 1893: Gabled hall with its date in stucco at the apex, built by the Plymouth Brethren; linked by a concrete brick porch to a smaller hall alongside dated 1955.

Nos.6-18: *c.*1840: Terrace of single-storey random basalt rubble houses with blockbonded brick dressings to opes, with the end house at no.18 rendered, and presumably 'riz' about 1900. Most houses considerably altered with Velux windows, tiled window surrounds, new windows or doors, but the character of the little houses is still apparent and would merit restoration. They may have been built originally as coastguard cottages.   *110*

Former Coastguard Station: *c.*1880: Terrace of two-storey stucco houses with hipped roof set back behind stone-walled garden, now converted into houses.

Nos.42-52: Hatfield Terrace: *c.*1885: Terrace similar to nos.47-59, with two-storey canted bays, dentilled cornice and wallhead dormers with apex boards.

No.54: Two-storey double-fronted stucco house with Dutch gables with ball finials, on road down to the seafront.

Holy Bridge: See Ballyholme Esplanade.

J

JORDAN AVENUE
Short street off Bloomfield Road, linking to Roslyn Avenue, laid out about 1935.

K

KENSINGTON PARK
L-shaped street of detached houses, mostly two-storey and often hipped, from lower Downshire Road to Maxwell Road. Laid out on the site of former clay pits around 1930, the street was originally given the more imaginative name of *Westward Ho!*, but had become Kensington Park by 1935. Nos.15, 22 and 28 have hipped roofs with scalloped Westmoreland slates.
See *Hogg 110*.

Kerrsland Terrace: See 20-36 Seacliff Road.

KILLAIRE AVENUE, Carnalea

No.2: *c.*1990: White smooth-rendered house with monopitch porch, garage, chimneys, and spiky gargoyle-finials to other projections.

Ailsa Lodge Nursing Home: *c.*1880, formerly *Craigview*: Large two-storey stucco house with quoins, corbelled chimneys and dentilled cornice at bays; good entrance pillars and gates. It was occupied in 1885 by one Henry McNeill. Recently much altered and extended, with plastic windows.

Bridge House: *c.*1874, formerly *Elsinore*: Two-storey stucco house with hipped roof, bracketed cornice, lugged segmental-headed windows; central porch in the five-bay front, with Tuscan pilasters, dentilled moulding and balustraded top; ground floor channeled; bow window towards sea. When Welch photographed it about 1920, the interior was very grandly laid out, with scagliola entrance hall and spacious staircase hall with lantern over. Built for Daniel Joseph Jaffé (father of Sir Otto Jaffé) whose initials DJJ remain on the frosted-glass inner door, it was occupied from about 1885 by F R Lepper, a director of the Ulster Bank, who apparently used Harland & Wolff craftsmen to install joinery work and (now gone) relief frescoes of reclining maidens à la Albert Moore. The present name of the house presumably derives from the iron pedestrian bridge over the railway giving access to the seashore.
See *Welch; Young p.624.*

No.8a: Hamlet Hill: *c.*1874: Gate-Lodge to Elsinore, picking up the Shakespearian reference; single-storey stucco building with hipped roof, central chimney-stack, bracketed eaves windows in moulded surrounds with keystones, beside the stone gate pillars of Elsinore.
See *Dean p.75.*

KILLAIRE PARK, Carnalea

From Crawfordsburn Road to Killaire Avenue, with a number of interesting recent houses.

KILLAIRE ROAD, Carnalea

A private road developed from the grounds of Killaire House.

The Fort: *c.*1890: Picturesque two-storey house with flint pebbles pressed into rendering; cast iron conservatory and entrance porch setting off very varied elevations; balustrading over courtyard wall.

Ardkeen: 1904, by Vincent Craig: Irregular two- and two-and-a-half storey Arts and Crafts house designed for William Murphy of Murphy and Stevenson, linen merchants. An audacious design mixing roughcast with red brick and dressed ochre-coloured stone in a mêlée of Dutch gables, stained glass and deep eaves, pivoting on an extraordinary circular corner window on the SW of the ground floor, over which the first-floor corner is nervously cantilevered.

Hamlet Hill, Killaire Avenue: the simple hipped-roof lodge to the former Elsinore, with bracketed eaves and keystones in the moulded window surrounds. (Peter O Marlow).

Ardkeen, Killaire Road: Vincent Craig's picturesque Arts and Crafts house of 1904 has Dutch gables, roughcast walls and an extraordinary nearly-circular corner window. (Peter O Marlow).

Gate lodge to Killaire House, Killaire Road: an unusual asymmetrical design, somewhat blandified in the course of recent renovations - the bay window previously had alternate mullions and vertical glazing bars. (Peter O Marlow).

Killaire House: a very fine ashlar stone house of about 1880 with fretted apex boards, kneelers supporting the eaves, deep bow windows and double-hung sash windows. (Peter O Marlow).

During the eighties and nineties this was the home of Roy Bradford, politician and author.
See *BT 20 Jun 1996*.

Killaire House: *c*.1880: Substantial two-storey house in dressed stone ashlar with end bays set slightly forward carrying fretted apex boards on kneelers. A semicircular bay on the ground floor with five elegant segmental-headed windows to the left of the urn-topped porch which has fielded Corinthian pilasters. Matching gatelodge and outbuildings. Originally known as *Ballykillare House*, from the name of the local townland. Occupied in 1881 by Captain A M Henderson whose monogram is in the frosted glass of the inner door - he was probably its first occupant - and in 1888 by S C Davidson, of the Sirocco Engineering Works, who later moved to Seacourt (see 120 Princetown Road).
See *Dean p.62*.

Kingsland Park: See Seacliff Road.

KING'S PLACE
Narrow lane rising steeply from Southwell Street to join The Vennel, with a short link to King Street that had a terrace of four houses on it; there were also the two ARGYLE COTTAGES that faced King Street. It was also known as *West Place*. Demolished about 1980.
See *Seyers p.2*.

KING STREET
Originally known as *Souter's* (ie Shoemaker's) *Row* or *Sooty Raw*, and about 1880 as *West Street*, the street was developed before 1833. The present name had become general by 1910. The existing buildings date mostly from the late 19th century and have little individual architectural interest, but this is one of the last residential streets in the centre of Bangor, and currently under threat of redevelopment.
See *BNL 6 Sept 1995; Spectator 22 June 1995, 7 Sept 1995*.

Nos.5-11: Penguin, house, The Winning Post: *c*.1900: Terrace of two-and-a-half storey red brick houses with glazed-sided attic dormers above deep boldly corbelled eaves. Only no.7 is still original, with its bootscraper, very wide ground floor window, vertically-divided first floor sashes and tall brick chimneys.

Nos.37-43: *c*.1900: Stepped terrace of smooth-rendered houses with steps up to recessed front doors, presumably reflecting the need to avoid obstructing the pavement of the narrow street. No.39 has the original four-pane sashes.

Nos.20-34: *c*.1912: A simple terrace of two-storey red brick houses, but several still in original condition with vertically-divided sash windows and four-panel doors. In 1906, *Mr Edward Cooney's Temple* occupied part of this ground, but it is not clear whether it was a mission hall, a gospel tent, or something altogether more pagan.

KINNEGAR: see *Queen's Parade.*

Knightsbridge: See 66-72 Seacliff Road.

KNOCKMORE PARK
Road from Ranfurly Avenue to Maxwell Road, laid out between 1903 and 1921, but most of houses inter-war in date.

No.1: Cressington: *c.*1920: Roughcast Mediterranean villa with a proliferation of palm trees; it used to have large eagles on each gate pillar.

Nos.3-17: *c.*1930: Four pairs of semis with half-timbered mutual gables, each one slightly different but many still with sash windows.

Nos.19 and 21: *c.*1925: Two of the original Ulster haciendas, two-storey rendered villas with shallow hipped pantiled roofs with deep eaves, external shutters to first floor windows, Art Deco central features and sweeping side wings: a touch of Hollywood.

No.31: *c.*1920: Two-storey roughcast house with hipped bays at ground floor, and tripartite windows over; double-hung windows with nine-pane upper sashes; roundels in gables. Pillars capped with rosemary-tiled gables.

L

LANCASTER AVENUE
Short cul-de-sac off Manse Road with roughcast two-storey houses, mostly semis built about 1930.

Landerville Crescent: See 36-54 Dufferin Avenue.

Linwood Terrace: See 10-20 Alfred Street.

LISNABREEN WALK
Cul-de-sac off Skipperstone Road.

Lisnabreen Presbyterian Church: *c.*1990, by Knox & Markwell: Irregularly planned church in warm brown brick, presenting a concrete-pantiled roof and gable to the front, with a heavily recessed round-headed door in the centre.

Little Clandeboye: See Main Street, Conlig.

LONG HOLE: see *Seacliff Road.*

LORELEI
Lorelei: *c*.1890-1900: Three-storey terrace of six stucco houses accessed from a lane between 82 and 88 Princetown Road, but looking directly over Bangor Bay. Paired three-storey bow windows; iron balconies at nos.3 and 4, which were built about a decade before nos.1, 2, 5 and 6 were added for Samuel Crosbie by Young & Mackenzie. Sometimes spelt *Lorely*, latterly most of the terrace has been the *Tedworth Hotel*. Following a recent planning appeal decision, demolition of all but the facade is likely to take place in the near future.
See *App 15; Bldgs at Risk.vol 3 p.52; Eakin; IB 15 Aug 1898; Lawrence 9538; Spectator 18 Sep 1997.*

LOWRY'S LANE
Lane off Crawfordsburn Road, formerly leading to *Lowry's Farm*, a two-storey five bay rendered house built before 1830 which was demolished about 1980 and is now a housing development.

LOWRY HILL, Carnalea
Lane off Crawfordsburn Road originally leading to two houses known as *Belle Vue Villas*.

Ballylogue House, formerly *Bellevue House*: *c*.1880: Two-storey stucco house with half-hipped roof and moulded surrounds to windows.

Belle Vue Villas: *c*.1880: Two-storey hipped-roof semi-villas built by Mr Lowry of the Hill for one of his sisters. One house now *Deepwell House*.
See *Spectator 7 Aug 1997.*

Luskinyarrow Terrace: See 4-14 Bingham Street.

LYLE ROAD
Road parallel to Ballyholme Esplanade from Waverley Drive to Sandhurst Park, in the coastal tradition of building more comfortable houses sheltered behind those on the seafront. Laid out before 1903, but in the event it was never fully developed and has become more of a service road, building occurring on the new spine roads developed towards the Groomsport Road.

No.80: McCloy Fold: *c*.1990: Smaller than most sheltered housing schemes, and discreetly tucked away in the middle of a residential area.

M

MAIN STREET
The chief shopping street of the town, divisible into upper and lower Main

Lorelei: a particularly bold example of the late Victorian terraces overlooking Bangor Bay. Built in the 1890s and currently under threat of demolition, the detail and massing of these terraces are very important. (Peter O Marlow).

53-55 Main Street: among the urban vernacular of Bangor's main streets, Gilbey's building (here photographed in 1968) was an unusually sophisticated building with the facade recessed behind detached columns. (H A Patton).

Street, the former a level road extending from the railway station to the junction of Hamilton Road, the latter a steep hill thence to the sea. The street was established by the time of the Raven map in 1625, though it has altered very considerably in detail. The Parliamentary Gazetteer of 1844 describes Main Street as "spacious, somewhat neatly edificed, partly winged with alleys and brief subordinate streets" and as being "the chief seat of the local trade". Until two severe car bomb attacks in the early 1970s, the street was largely late 19th century in character, but several fine buildings in the lower Main Street were destroyed at that time, and planning control since has left something to be desired. A 200lb car bomb in lower Main Street on 21 October 1992 and another bomb in upper Main Street on 7 March 1993 caused further damage.
See *BT 1 Apr 1974; Eakin; Lawrence 2360, 2366, 3875, 3878, 4734, C6016, C6018, C2357, C2854; Parl Gaz p.214; Seyers pp.2-3.*

Nos.1-5: Kentucky Fried Chicken, Travel Care, Jazz Hair: *c*.1890: Two-storey six bay building with stucco quoins and recent tiled roof. This was originally the *Eagle Hotel*, later the *Yachtsman Hotel*.
See *Lawrence 2857.*

Nos.7-19: The Flagship Centre: 1992-93, by Ostick & Williams for Farrans Construction: Symmetrical largely glazed building with a timber-framed pediment on which a plastic owl sits to scare real birds. Behind the frontage runs a corridor of shops with most of the charm of an airport lounge. The Co-op Superstore that was erected on the site in 1965 was partly retained in the new frontage to Main Street. In the 19th century, no.19 was the POST OFFICE, run by John Matthews; when he died his son John took over. Being rather busy running a corn mill, a butcher's shop, a grocers and a bakery as well as the post office, he neglected to inform the authorities, who "never knew any difference".
See *Lawrence 2857; Perspective Mar 1994 pp.41-45; Seyers p.29; Spectator 16 Jun 1956, 22 May 1964, 3 Sep 1992, 12 Aug 1993; UA Jan 1994 pp.68-71.*

Nos.21-23: Shirt Centre: *c*.1880: Two-storey three bay re-rendered building; unusual modern shopfront of timber and steel.

Nos.25-27: Johnston, Benetton: *c*.1906: Three-storey four bay stucco building with modern shops; damaged by bomb in October 1992.

Nos.29-33: Easons: *c*.1995: Post-modern building in reconstructed stone with tiled ground floor; central bay recessed and roof over supported by two fairly stainless steel columns. The previous BUILDING on the site was a two-storey six bay stucco building of *c*.1890, with an attractive first floor of plain sash windows in moulded openings decorated with flowers and stars, framed by a corbelled cornice and stucco quoins; originally it had domestic doors in round-headed opes and a carriageway entrance alongside the single shop. The butcher's shop until recently at no.31 had been founded by the Bowman family (who also owned the brickworks at what is now Ward Park) in 1820.
See *Lawrence NS4302; Seyers; Spectator 15 Sept 1961; Wilson p.55.*

MAIN STREET

Trinity Presbyterian Church: 1887-88, by S P Close: The second Presbyterian congregation moved here from Brunswick Road in 1889, when the Marquess of Dufferin & Ava, Rev William Clarke, J B Houston and A Sharman Crawford each laid memorial stones on 28 September. Messrs H Laverty & Son were the builders of the "New Presbyterian Church and schools", which are built of very shallow coursed and rusticated Yorkshire stone of a warm honey colour, giving a stripey almost brick-like effect. The frontage is irregular, and roughly Early English in style, with the main gable containing a large triple lancet window flanked by heavy buttresses. The entrance is at the southern porch, while a squat buttressed octagonal turret forms the NW corner. A lecture hall was built in 1894 and a Minor Hall in 1931. The matching stone wall to the street has a modest lych-gate at the N end. The building formerly on the site of the church was known as BOWMAN'S HOUSE, and included modest stucco houses set back from the present building line with well-worn stone steps up to them.
See *IB 15 Mar 1887, 15 Aug 1888; NDH 27 Jul 1888; Presb Hist pp.113-15; Reid; Spectator 5 Sep 1931, 24 Feb 1961; Wilson pp.54-55.*

No.35: Dorothy Perkins: *c*.1965: A square box with plain strip of cement brick above clerestory lights. This was the location of Bangor's first cinema, the PICTURE HOUSE, which opened in April 1912. It had an entrance hall decorated with anaglypta and an "exceedingly artistic aspect of electric fancy lamps from each archway" which was much admired. The "cinematograph hall" itself could hold up to five hundred people watching a "steady and almost flickerless" picture. In its first week of operation, it showed film of "The Great Unionist Demonstration" at Balmoral taken "exclusively" for the cinema, enabling patrons to be advised to "Come and see yourself on our screen." By the end of the month it was showing film of the building of "the ill-fated S S Titanic". The Picture House moved to Quay Street some years later, but re-opened in 1928 as the *Adelphi Kinema*. The newly opened cinema advertised that, in order to counter germs, the theatre would be flooded with daylight during the day. In the time just before the talkies, it invested in a musical installation containing "electricity of very high voltage and low frequency" which was startlingly realistic. In June 1957 the *Tudor Cinema* opened in "premises formerly known as the New Theatre", with the entrance strangely decorated to look like a "very old stone passage" with the walls hung with shields and coats of arms, and imitation slit windows painted on either side of the screen. Ironically, a television shop with "palatial new showrooms" took over the site in 1960.
See *Spectator 5, 12 and 26 Apr 1912, 27 Oct 1928, 29 June 1957, 9 Dec 1960; Wilson p.55.*

Nos.37-47: Little Imp, Pavilion, Pound City, Café Wills, Granada: 1889 for James Neill: Group of four related gable-fronted buildings with modern shopfronts. Gables are alternately three-and-a-half storey with pitched-roof dormers and four-storey, the top windows all being round-headed, and each

building with rusticated pilasters. The gables used to terminate with ball finials enjoyed by seagulls. In 1908, the *Spectator* noted that "Where a year ago plain unvarnished shop fronts were the rule, we now have fine ornamental fronts... French-polished or varnished", and picked out the premises of Messrs A Harper & Co, who were at no.41, as "a very good example of city style... remodelled in the leading style by Mr Alexander Macrae."
See *NDH 22 Feb 1889; Spectator 31 Aug 1908.*

Nos.49-63: Heart Foundation, Bingham Mall, Winemark, JJB Sports, Halifax: *c.*1975: A collection of new shops built after the 1972 car bombs. Nos.53-55 were formerly W & A GILBEY LTD's fine Victorian premises, with colonnaded upper storeys and elegant shop, while nos.57-63 were three-storey stucco buildings. The replacements tend to the brisk, bright and cheap. *118*

Nos.65-65a: Simpson's Sewing Shop, Espionage: *c.*1905: Three-storey three bay stucco building with two modern ground floor shops. Upper windows altered, but in original opes with moulded surrounds; decorated pilasters framing the elevation and remains of the Edwardian shopfront installed by Sam Nelson, ironmonger and oil merchant.
See *Spectator Directory 1906.*

No.67: Progressive Building Society: 1984: The rather attractive three-storey SHOP with bargeboard formerly on this site was demolished in 1983, and has been replaced by a two-storey brown brick building with an asbestos slate roof; presumably a sample of progressive building.
See *Spectator Directory 1906.*

Nos.69-73: Woodsides: *c.*1970 for Dunne's Stores: A very bland supermarket building with chamfered fibreglass panels above the shop entrance. Site formerly occupied by a pair of three-storey Victorian HOUSES that became the *Downshire Temperance Hotel* around 1900, to which shopfronts and first-floor oriels were added around 1920. In 1910 the hotel's proprietor advertised that he had a hall capable of seating two hundred people comfortably.
See *Eakin; 1910 Directory p.1536.*

No.75: Ulster Bank: dated 1920, by James A Hanna: Three-storey red brick building with stucco pediments, pilasters, and ground floor. The central bay is set slightly forward, with the central pediment and cornice ornamented with shallow dentils borne on plain pilasters decorated with wreaths and lettering discreetly placed on the cornice band. The side elevation has baroque scrolls supporting the gablet and chimney above, while at first-floor level is a shallow decorated balcony supported by a gryphon. Over the front door, lions with folded paws form corbels to the porch, which contains the bank's coat of arms. About 1990 the bank proposed to demolish the building (apparently because it was thought to be the easiest way to move the safe), but this met with considerable public outcry and in the event it has simply been extended discreetly to the rear. This building originally incorporated a house for the manager on the upper floors. It replaced the BELFAST HOTEL *iii, 123*

MAIN STREET

which had been on the site in the 1860s, with a public weighbridge between it and the courthouse. The previous building also served around the turn of the century as the *Adelaide Tea Rooms* and as the *Reading Rooms* before the erection of the Carnegie Library. Hanna was a prominent architect in the province in the early years of the century; a fastidious dresser who still wore a silk hat in the 1920s, he was known in his family as Aunt James.
See *App 882; Eakin; Lawrence 2360, 2854, 3875, 4734, C2357; Ulster Bank p.168.*

iii, 124 **No.77: Northern Bank:** *c.*1820, formerly *The Court House*: This was Bangor's Market House, built by Lord Bangor and Col Ward some time between 1770 and 1820. Dubordieu, writing in 1802, records that wheat, barley, raw hides and calves shins were bought up in Bangor for the merchants of Belfast, but does not record whether the transactions took place in the market building; the 1824 Directory in fact records "a Market House without the usual accompaniment of a market", and the Ordnance Survey Memoirs describe it as "a small building of recent erection and plain and unfinished appearance". The date usually given for the building is about 1780, based on a 1777 account of an *Assembly Room* fifty feet by twenty-four with "beautifully stucco'd" ceiling and light blue walls, and an account of the American Paul Jones shelling Bangor and firing two thatched houses with "a red hot shot" which hit the "roof of the new Market House". However these reports seem to refer to a different building - the Irish Corporation Commissioners Report in 1834 says that "The Lords of the Manor had lately built a market-house", adding that "In September 1813, an order of the Corporation was made that the provost should provide a proper place for building a market; but we do not find anything done upon it", which would suggest a date about 1820.

The building is of stucco, two-storey and five bays wide with quoins and a balustraded parapet. The three central bays are set forward under a pedimented gable containing a clock, and the first-floor windows are surrounded by moulded architraves broken by Gibbsian cubes. The architect played a Mannerist joke by recessing the central windows and setting the architraves of the outer ones forward. The ground floor was originally arcaded "with iron gates where windows are, beams and scales" in the 1860s, but was filled in when the building became *Male Female & Infant National Schools* (commonly known as the *Ward School*) about 1890, and the ground floor fenestration was altered again more recently to put the entrance centrally rather than at the southern end. In the 1880s, the Bangor Musical Society organised concerts on the first floor.

Although still a dignified building, it is less imposing since it lost the octagonal cupola that used to grace the hipped roof, and chimneys that rose from the parapet on the south gable. In 1933, the building became the *Town Hall*, and in 1952 it became the premises of the Belfast Banking Company, now absorbed in the Northern Bank.
See *App 6741; Arch Surv p.395; Brett pp.63-64; Dubordieu p.225; Lawrence 2854, 2856, C2351, NS4406; Lowry p.lxxxvi; OS Mems pp.24-25; Seyers p.3; Spectator 9 Sep 1933, 27 Mar 1954; Wilson p.4.*

75 Main Street: when James Hanna designed this bank in 1920 the manager lived over the shop, so the upper floors were a house. A crucial building round which the two parts of Main Street pivot. (Peter O Marlow).

77 Main Street: originally the market house below and court house above, the ground floor had open arcades, and the roof was topped with a cupola till about 1950. (Peter O Marlow).

109 Main Street: the Art Deco premises of the Spectator newspaper, designed in 1924 by Gordon O'Neill. (Peter O Marlow).

68 Main Street: forget the gaping hole on the ground floor and enjoy the curly gable and curved corner. (John Gilbert).

No.79: Lyons House: Boots: by Robinson & McIlwaine, 1982: A rather aggressive red-brick department store which has replaced former hipped three-storey stucco BUILDINGS, which were known as *Warden's Corner* from the shop on the ground floor.
See *Lawrence NS4649, 2856; Wilson p.4.*

Nos.85-87: First Trust Bank: *c.*1995: Three-storey building of smooth red brick with raised skews to roof and ornamental string courses above ground floor and first floor.

No.89: Vacant: *c.*1890: Three-storey stucco building with modern ground floor shop and shallow first-floor oriel window; all windows segmental-headed. This is the surviving fragment of a building that used to include nos.85-87.
See *Lawrence NS4649, C6018.*

Nos.93-95: Vacant: *c.*1903: Three-storey four bay stucco building with modern shopfronts, which replaced earlier low two-storey stucco HOUSES.
See *Lawrence NS4649.*

No.97: Vacant: *c.*1890: This is the surviving bay of a larger building, with fine stucco decoration to the first floor oriel bow window, grouped round-headed windows at second floor and dentilled cornice with its corbelled-out end pediment. It was originally topped with balustrading and a small pediment.
See *Lawrence C6018, C2351.*

No.99: Menary's, vacant: *c.*1970: The *Bungy Jump Restaurant* occupies the site of the rest of the building from which no.97 survives; white mosaic with a glazed first floor. The portion to the south (*c.*1900) is an earlier three-storey three bay building with decorated stucco upper floors, which originally had oriel windows at first floor.
See *Lawrence C2351, C6018.*

No.109: Spectator Buildings: 1924, by Gordon O'Neill for D E Alexander: *124* A striking rather Art Deco building (replacing a low two-storey one) with Diocletian windows at first floor and a small pediment breaking through the cornice above the second floor carrying a wreath and the information "Established 1904"; this refers to the *Co Down Spectator* rather than to its premises (the paper was formerly based across the road at nos.124-126). The *Spectator*'s editor for many years, the Scot D E Alexander, said "I print for a living, but the Spectator is my hobby". The architect H A Patton, for many years Bangor's town planner, had his office here in the 1960s.
See *App 1581; Lawrence C6018; Spectator 6 June 1953.*

Nos.111-115: Cleaver Fulton & Rankin, Pizza Hut: 1984: Four bay brown brick building with projecting canopy. The previous BUILDING on the site was a trim two-storey stucco one with two small roof dormers.
See *App 7471; Lawrence C2351.*

Nos.117-121a: The Front Page, Going Places: *c.*1995: Two-storey rendered building with raised gables and reconstructed stone ground floor; gable to

MAIN STREET

Market Street. The architect Gordon O'Neill had his offices on the second floor of nos.119-121 around 1922.
See *Lawrence C2351*.

Nos.125-137: Tuki and Singing Kettle, Radius, Linleys, Robert Neill Coal: 1904-06 but largely rebuilt: Three-storey nine bay building with modern shopfronts. Before the building was bombed in March 1993, and subsequently largely demolished, the upper windows had moulded surrounds and there was a dentilled cornice. **No.139** (Home Interiors) similar, brick built. When this was originally constructed it was the new *post office*, complete with the postmaster's house and a yard with space for a bicycle and handcart. In 1920 nos.135-137 were a *Temperance Hotel*.
See *Lawrence 2854, C6018, C2351; Spectator 13 July 1906, 12 Aug 1993.*

No.143: Post Office: 1936, by T F O Rippingham of Ministry of Finance: Two-storey seven bay building in rustic brick with creamy stone plinth and steel casement windows. The heavy pantile roof has deeply overhanging eaves crushing a tiny clerestory on to the heavy cornice. The design is symmetrical, with central doorway, and additional smaller bays set back at each end; ground floor windows set in arched niches, first floor windows with steeply splayed soldier courses. Opened in 1936, with an Edward VIII letter box - one of only two post offices in the United Kingdom so graced - the Postmaster Surveyor told his guests that it had been designed with materials chosen to give "a quiet dignity allied with colourful restraint" to "a public which has a growing appreciation of beautiful things". The building replaced a TERRACE of two-storey stucco houses nestling into the demesne wall of Bangor Castle.
See *Hogg 29;* photo in NDHC shows previous building before demolition; *Spectator 29 Aug 1936; Wilson p.52.*

No.2: Dunkin Donuts: *c.*1895: Three-storey corner building with giant order pilasters and ornamental string course below second floor; octagonal corner turret. Stewart Aicken had a single-storey PUBLIC HOUSE here in the 1860s, which was redeveloped at the end of the 19th century. In the 1930s, as John Lynch's *Central Bar*, it was further altered by John McBride Neill, who may have added the bellcast roof to the corner oriel. Ground floor now modernised and in shops.
See *Crosbie p.10; Eakin; Spectator 22 Oct 1932; WAG 328; Wilson pp.8, 26.*

No.8: T Oscar Rollins: *c.*1930: Nondescript building altered nondescriptly.

No.12: Wesley Hall: 1891: Two-storey hall with gable to street and elegant eyebrow over main lancet window. Connected with Queen's Parade Methodist Church of the same date.

No.16: Woolwich: *c.*1900 and later alterations: Three-storey rendered building, partly rebuilt after bomb damage, without replacing details; formerly the *Rendezvous Restaurant*, it had shallow first-floor bow oriels and fluted pilasters to the shopfront.

143 Main Street: Rippingham's 1936 design for the new post office combines Neoclassical arched windows and brickwork with almost Mediterranean pantile roof and deep eaves. (Peter O Marlow).

Main Street: a rare photograph of a house, probably dating from the 18th century, near the site of the present post office; with limewashed stone walls and roof of local Tullycavey slates, it was demolished about 1910. (NDHC).

82 Main Street: Millar & Symes' 1934 Bank of Ireland still retains some original shopfront details. The dark brick is contrasted effectively with the crinkly skyline and central clock tower. (Peter O Marlow).

80 Main Street: the most recent of the four bank buildings at the top of Main Street, this was built for the Allied Irish Bank in 1969, replacing the old Imperial Hotel with an asymmetrical Brutalist structure. (Peter O Marlow).

Nos.18-20: Halifax: *c.*1975: Irregular building in bright red bleeding brickwork with metal shop fronts and fake mansard roof. Nos.18-20 were previously a plain three-storey stucco BUILDING, including Lennon's splendid fruit shop which was blasted in 1972.

Nos.22-30: F W Woolworth & Co.Ltd: *c.*1955: Originally opened in Bangor as a "3d. and 6d. store" in 1930, the present building is very uninteresting - a two-storey flat roofed horizontal building with a ground floor of tiles and glass, and pebbledash above.
See *Spectator 21 Feb 1956.*

No.34: N I Hospice: *c.*1979: Two-storey flat-roofed brick building.
See *Lawrence 2366, C6016.*

No.36: Oxfam: *c.*1890 but much altered: Two-storey refacing of stucco building formerly occupying nos.34 and 36, which formerly had quoins and elaborate stucco relief borders to the first floor windows.
See *Lawrence 2366, C6016.*

Nos.40-42: Stewart Miller: 1984: Building in dark brown brick with strip window at first floor; replacing the former pair of two-and-a-half storey SHOPS with gables and shallow first floor oriels.
See *Lawrence 2366, C6016.*

Nos.44-52: Wellworth's: *c.*1965: Supermarket frontage originally derived from the Woolworth pattern (see nos.22-30) but reclad *c.*1993 with three bleak curtain-walled gables set forward from white panels. [Closed 1999].
See *Lawrence 2366, C6016.*

Nos.54-56: Birthdays: *c.*1993: Two-storey purplish-red brick building with gabled bay set forward and side piers in black brick. Replaced a pair of gableted brick BUILDINGS of *c.*1920 (badly damaged in the 1992 car bomb), which housed Scott's fish shop (he used his own boats to get fish on to his marble slabs, as most of the local catch went straight to Belfast), and in the 1970s the offices of architect Stanley Devon.
See *Lawrence 2366, C6016; Spectator 22 Oct 1992.*

Nos.58-60: New Look: *c.*1980: Dark-brown brick building with polygonal first floor window; replacing the former delightful two-storey stucco drapery SHOP of Margaret Campbell, which had marbled lettering and a row of four round-headed first floor windows.
See *Lawrence 2366, C6016.*

Nos.62-64: Barratt's, Abbey National: *c.*1960: Pair of two-storey flat-roofed shop units of frame construction. The site was originally part of the building surviving at no.66.

No.66: West End House: H Samuel: 1884: Three-storey three bay stucco building with giant-order panelled pilasters and moulded surrounds to upper windows; dated on gable below stump of a chimney. Part of what was originally a seven-bay building. Until recently it was the premises of Hugh Furey's, the original builders, drapers turned publicans: the entrance to their

bar was decorated with a mosaic gable mural of *c*.1960 depicting sun, fishes and birds.
See *Crosbie p.7; Lawrence C6016, 4734; WAG 3102.*

124 **No.68: Scrabo House:** Clark's Shoes: *c*.1890: Ornate two-and-a-half storey stucco building with chamfered corner and frilly Dutch gable to King Street; deeply corbelled chimney, ornamental clay ridge, dentilled cornice. The shopfront accommodates the curved corner, but an unfortunate box dormer replaces an earlier gabled one, and the windows have recently been changed to plastic.
See *Crosbie p.7; WAG 3102.*

Nos.70-72: Etam: *c*.1930, by Thomas Callender for Smyth & McClure: Three-storey inter-war block, flat-roofed with some very basic faintly Art Deco ornament. In the 1950s, most grocery shops in Bangor expected their customers to sit down and dictate their order to the assistant behind the counter, but Smyth & McClure's had revolutionary open shelving.
See *App 4306.*

No.74: Excel: Three-storey flat-roofed building, the top storey added to a structure originally similar to no.68.

No.76: Nationwide: *c*.1955: Two-storey building in rustic brick.

No.78: David Mawhinney: *c*.1930: Two-storey four bay building with full-length aluminium shopfront. Sergeant Buchanan, the last Town Sergeant of Bangor, lived in the house previously on this site.
See *Lawrence 4734; Spectator 5 Sep 1931.*

128 **No.80:** Clinton Cards: 1968-69, by Shanks & Leighton for the Allied Irish Bank: Three-storey concrete building faced with white mosaic; of asymmetrical design with oversailing second floor pierced by groups of lancet windows, and the ground floor infilled with dark brick. The contractor was William Dowling Ltd. Quite a striking, if fairly uncompromising, design, the site was formerly occupied by the gabled stucco IMPERIAL HOTEL of *c*.1880, which offered the best meals in the town in the 1930s and was known as *The Widow's* from its owner, Widow Morgan. In the summer of 1904, when wet weather prevented their performing at the bandstand, Miss Ella Rose's Golden Star Concert Party would perform in the *Imperial Assembly Rooms* adjoining the Imperial Hotel.
See *Crosbie p.6; Lawrence 2360, 3875, 4734; Seyers p.29; Spectator 17 Jun 1904; WAG 398A: Wilson p.5.*

128 **Bank of Ireland:** 1935-37, by Millar & Symes: Three-storey seven bay building in dark rustic brick with rendered ground floor, central doorway in indented opening, original shop openings, and small-pane upper windows. Zigzag balconette to central first-floor window, deep plaster cornice with a crinkly top and squat crinkly-topped central tower, like a baby skyscraper. The shops were originally occupied by the Belfast Co-operative Society, and part of the original fenestration can still be seen on Central Avenue. Millar &

MAIN STREET

Symes were Dublin architects who worked regularly for the bank. This was the site up till the middle of the 19th century of the BLACK HOLE, a place of confinement to which the Provost of Bangor could send disturbers of the peace. Its walls were from three to four feet thick, and so strongly built that its demolition required blasting. It was removed to make way for the BANGOR ENDOWED SCHOOL, a charming two-storey three bay stucco building with mullioned windows under label mouldings, and drop finials at the corners of its projecting eaves. It was built in 1856, and taught around fifty boys under the terms of the will of Hon Robert Ward. When the school moved to *College Avenue*, the old building served as the *Town Hall* from 1900 to 1933.

87

See *App 4010; Seyers; Spectator 5 Sep 1931, 3 April 1937, 10 Jul 1964; Wilson p.78.*

Nos.84-88: Mortgage Shop, Action Cancer, Ulster Property Sales: dated 1875: Terrace of two-and-a-half storey buildings with modern ground floor shops, with skewed gable and truncated corner. Frilly bargeboards to dormers and first floor mouldings, almostly entirely removed in recent years, gave this group strong character. Proudly dated in stucco shield 'AD 1875'.
See *Lawrence C6018.*

No.90: Wrights: *c.*1968, by H A Patton: Three-storey blue-tiled and flat-roofed building with five tall projecting rectangular oriels at first floor; originally part of the terrace nos.84-88.
See *Lawrence C6018.*

No.96: Heatherlea: *c.*1965: Bland two-storey building fusing timber and aluminium; originally part of the same terrace as no.100.
See *Lawrence C6018.*

No.100: John Neill & Sons: *c.*1880: Two-storey two bay stucco building with modern shopfront; gatepillar at gable for the adjoining church.
See *Lawrence C6018, C2351.*

First Bangor Presbyterian Church: 1831, with later additions: The main body of the church has a rounded front built of basalt rubble in two storeys, with a three-stage tower and steeple added in 1881 in unmatching "scrabo freestone with Scotch red sandstone dressings", and a basalt portico. The body of the church has segmental-headed windows at ground level below round-arched first floor windows, with a plain broad sandstone cornice band at the eaves. The tower, which has Early English detailing, rises from a basalt base above the original vestibule to a sandstone stage with two lancet windows on each face and an oculus breaking the string course below the belfry, which has louvred-arched windows set between pilasters and is topped by balustrading and corner finials. The slender white spire is octagonal with aedicules at three stages. The portico (added in 1928) faithfully reproduces the appearance of the original elevation, with giant-order pilasters supporting the broad band cornice and separating the strange squat semicircular windows with pendant labels at first-floor level. There are three doors, the central one of which has a triangular stone pediment.

133

Inside the vestibule one faces the original front of the church to which a double staircase was added giving access to the gallery. The auditorium is D-shaped, with the box pews following the plan of the church and facing the highly polished arcaded pulpit and the organ with its primrose-painted pipes disposed in a segmental-arched proscenium. The gallery is deep, with sheeted fronts and colonettes, closely bracketed and supported on Doric columns, and the ceiling is coffered and sheeted. The organ was presented in 1908. The stained glass includes a window to W G Lyttle the writer (best known for his novel *Betsy Gray*), and another with a small medallion showing the church before the addition of the portico. There are marble monuments to two former ministers, Rev J C McCullough (1857-78) and Rev Alexander Patton (installed 1879); the latter records "An eloquent preacher, a faithful pastor, a zealous worker, a wise counsellor and a sincere friend", and carries a discreet hourglass on the keystone.

In 1623, one Robert Blair came to Ireland from Ayrshire to take up the living of the Parish Church - on the face of it an odd appointment, since he held strong Presbyterian views; however, these probably coincided with those of his patron (Sir James Hamilton had been a secret agent for James VI, who hated bishops). He was obviously as reluctant a visitor to Ulster as any in recent decades: "When I landed" he wrote, "some men parting from their cups, and all things smelling of a root called rampions, my prejudice was confirmed against the land. But next day travelling towards Bangor, I met unexpectedly with so sweet a peace and so great a joy as I behoved to look thereon as my welcome hither". He stayed on in Bangor, but proved too heretical to retain his episcopal appointment, and he was suspended in 1634. However, he did have a large following in the town, which under the leadership of Gilbert Ramsay became Bangor's first Presbyterian congregation, and they went on to build a church on Fisher Hill (now Victoria Road) about 1660. This was destroyed in 1669 on the orders of Alice Countess of Clanbrassil, but rebuilt on the same site in 1685. In 1741 the "Old Meeting House" was rented by the congregation at the corner of Ballymagee Street (High Street) and Quay Street, for one shilling per annum, and this served till the erection of the "new and beautiful building" described by Lewis in 1837.

It was on 1 June 1831 that William Sharman Crawford laid the foundation stone of the present church, a fine day when "a very great number of people attended" and, as the Belfast Newsletter's reporter ruefully noted, "were addressed at great length by Rev Hugh Woods", who was minister from 1808 till his retirement in 1856. The basalt church, which consisted of the present auditorium with an elevation of wrought sandstone to the Main Street and cost £2400, did not acquire its spire until after the installation in 1879 of Rev Alexander Patton, great-grandfather of the present writer, who energetically set about raising funds, commissioned two designs for the spire, and by the looks of it chose the less appropriate one! Sandy McFerran of Bangor was the contractor for this bold addition, and possibly for the small north and south porches added about this time. The *Guild Hall* and *Schools* to the rear

First Presbyterian Church, Main Street: This photograph taken about 1900 shows the original church portico and the newly added spire. In the background can be seen the developing Princetown area. (Welch Collection).

MAIN STREET

were erected in 1894. In 1928 the front portico was extended by an extra bay, and a plaque inside records that "This Vestibule and Stairways, also the Emergency Exits" were erected by James Thomson; the clock was also presented about this time. A recent extension has been added to the south. The original manse of the church was built in 1867 on the site of the present shopping complex to the south of the church, but it was latterly a draper's shop before it was demolished in 1972.

The church's spire is a very prominent landmark in the town, forming a counterpoint with the darker spire of St Comgall's, and although the building is set back from the building line of the street, the weeping ash and front railings nicely punctuate the streetscape.

See *Arch Surv pp.340-341; BHS I pp.22-24, II pp.18-22, III pp.46-51; Crosbie p.5; First Bangor Magazine Dec 1958; Lawrence 3883; Lowry pp.38-39; Milligan pp 10-11; Morton p.18; OS Mems p.24; Presb Hist pp.111-13; PRONI D.2194.xi.26; Seyers p.35; Spectator 17 April 1908; WAG 395; Welch 35; Wilson p.58; Wilson Wm passim; Young p.143.*

Nos.102a-112: *c.*1975: Cancer Research, Surf Mountain, Roseby's, Geeves, Wheatear: Two-storey block of shops under first floor windows with projecting plastic panels; developed by First Bangor Presbyterian Church on the site of former MANSE AND SHOPS which were variously two-storey with running label mouldings and round-headed doors; single-storey; and three-storey with segmental-headed windows.

See *Crosbie p.4; Eakin; Lawrence NS4478, C6018, C2351, 2854; Spectator 24 Feb 1911, 25 Mar 1933; WAG 398.*

iii **Nos.114-126: Menary's:** 1997-98, by Robinson & McIlwaine for Ulster Estates Ltd: Two-storey rendered building with large first floor windows angled down towards the street, with three metal-roofed glazed gables at one end. At the turn of the century this was developed with three-storey houses with attractive shallow castellated bow-oriels, known as ASHLEY BUILDINGS. Nos.124-126 were redeveloped *c.*1970, for ROBINSON & CLEAVER'S, who occupied a three-storey brick box clad with fibreglass panels and glass ground floor; nos.114-118 were similarly redeveloped by Ulster Estates a couple of years later, but both buildings were damaged by a car bomb in 1993, and the present building is a replacement of nos.114-122 and refacing of nos.124-126, with a large porthole window added on the southern gable.

See *BT 18 Sep 1998; Crosbie p.4; Eakin; Lawrence C2351, 2854; Perspective Sep 1998 pp.25-28; Spectator 25 Sep 1997; WAG 398.*

No.132: Ava Hotel: *c.*1840: Two-storey five bay stucco building with quoins; a recent shelter at front door replaces an earlier enclosed porch, and there is a sort of car port extension at the rear to the "Dufferin Rooms", fortunately screened by a well-placed tree. An interesting outdoor museum of old advertising signs was recently created beside the curious oriel-windowed building at the top of Dufferin Avenue. Built as a house by Dr Russell who ran the dispensary in Catherine Place, it became the *Railway Hotel c.*1870,

then a private house again before becoming the Ava Hotel.
See *Apps 1064, 1100, 1151; Hogg 24; Seyers p.1; Spectator 11 April 1996; Lawrence 9554.*

MAIN STREET, Conlig

The road from Bangor to Newtownards originally included the narrow main street of Conlig but it is now bypassed and has a separate entity.

No.29: *c.*1860: Pretty gloss-painted cottage with narrow segmental-headed windows, set behind trees and tall bushes.

Conlig Presbyterian Church: *c.*1848: Spiky church in slightly squared random rubble and ashlar dressings which looks eastward from its hillside location. Gabled front with finialed buttresses and slender colonettes to the door and three lancet windows above; square belfry with finials, and clockface in base; round-headed windows with traceried tops along the body of the church. Linked by a modern hall to the old **School** (established in 1833) just north of the church. This is also of random rubble stonework, with a steep gable containing a quatrefoil over two narrow lancet windows, and low stone wall. The grounds of the church are mostly car park, with a solitary but important beech tree trying to look cheerful in the midst of the tarmac.
See *BHS I pp.19-20; OS Mems p.22; Spectator 5 June 1948.*

Gate Lodge: Rubble basalt lodge with ashlar quoins; projecting eaves with ornamental bargeboard; windows slightly pointed with heavy bossed mouldings over; projecting front porch and quadrant gate screen.

This was the lodge to LITTLE CLANDEBOYE (formerly *Conlig House*), a neo-Tudor mansion of *c.*1820 in bluestone rubble with sandstone dressing to oriels and turrets. It was situated on the edge of the Clandeboye estate and was at one time a dower house for the Blackwoods, though later occupied by William Pirrie, the shipbuilder. The interior included Gothic arched doors and a marble-floored hall. In the 1930s it was a mental hospital before becoming a particularly romantic ruin.
See *Dean p.82; Hogg H05/54/1-9.*

158

Lead Mines: The mining of lead at Conlig was first carried out about 1780, but failed because of flooding, and it was not revived till about 1827. More advanced technology then enabled it to thrive and employ up to four hundred workers at its peak, the lead ore being exported to Flint in north Wales for purification. Before closing in 1865, the industry had worked through six miles of underground shafts. The remaining structures are mostly ruinous and small in scale, but occupy dramatic countryside.
See *Eakin; McCutcheon p.231, pl.46.4; Spectator 15 Jun 1935, 3 Apr 1937.*

MAIN STREET, Crawfordsburn

At first sight a picturesque and unspoilt street of whitewashed cottages with black half-doors, and even a thatched cottage; but on closer inspection, the

MAIN STREET, CRAWFORDSBURN

thatch is only a small part of a massive hotel, and few of the doors and windows have much antiquity. There is a story of a monk "with a cowl over his head and a rope tied round his waist" walking through the village and disappearing through the walls of one of the cottages, but today the ghost might find the plastic doors and cavity walls offer more resistance.
See *Country Club p.69; Eakin.*

The Water Mill: *c.*1800, but extensively rebuilt: The pre-1830 water mill on this site which had been converted to flats in the 1960s was further altered *c.*1996. Two-storey smooth-rendered house set below the road, now with double garages as the main feature inside the automatic double gates.

No.1: Otira Cottage: *c.*1820: Two-storey smooth-rendered house with segmental-headed windows over irregularly spaced ground floor windows with very low cills. In terrace with nos.3-7.

Nos.3-7: *c.*1820: Two-storey smooth-rendered terrace slightly set back from, but in terrace with, no.1; all windows modernised, but with reasonable new shopfront at no.7.

No.11: *c.*1850: Two-storey smooth-rendered house with half-timbering including pigeon holes at apex of the gable towards the Old Inn. Four panel door with blocked corners to panels.

137 **No.25: The Old Inn:** Complex of smooth-rendered white-painted buildings of different ages, but mostly two-storey and post-war. The famous thatched portion which is single-storey with mullioned iron lattice windows is comparatively small and claims great antiquity. It is certainly 18th century in origin and may be much older (the Inn claims it dates from 1614 and was used by smugglers, and in 1798 by rebels), but has seen much alteration including the recent English-style reed thatch and scalloped ridge.
See *Country Club pp.17-22; Crosbie pp.58-59; Eakin; Lawrence 2848, 9557, 9558; Spectator 26 Oct 1935, 8 May 1937.*

No.2: Burn Cottage: *c.*1830: Single-storey roughcast cottage at the junction of Main Street and Cootehall Road, with porch on the gable and nasturtiums tumbling over the surrounding whitewashed stone wall. Monopitch extensions to rear.

No.2a: Sharman Lodge: 1914: Roughcast former community hall with battered pilasters and chimneys. Erected by Col Sharman-Crawford in memory of his son Terence who was killed in a motor-cycle accident at Aldershot in 1913. After a period as a schoolhouse, followed by dereliction, converted to private house by Bruce Crawford, 1979. Five large sycamore trees on a grassy slope below the lodge dominate the eastern end of the village.

Glen: At one end of a stone bridge which is modestly visible in the village is a path following *Crawford's Burn* and leading up through trees to the former MILL, of which a few rubble-stone walls survive. Beyond that the ground opens out into a park laid out around the former mill pond and the upper reaches of the burn.

The Old Inn, Crawfordsburn Main Street: most of the present Inn is quite recent in date, but the central thatched portion with its latticed windows is of considerable antiquity. (Peter O Marlow).

8 Main Street, Crawfordsburn: the former Cingalee Tea Gardens building displays a bizarre mixture of Tudor and Raj that fits surprisingly well into the black and white of Crawfordsburn. (Peter O Marlow).

MAIN STREET, CRAWFORDSBURN

No.6: The Cottage Studios: *c*.1800 but much altered: Single-storey roughcast bungalows with deep eaves and two-storey half-timbered gables at each end of the front elevation.

137 **No.8:** Crawfordsburn Country Club: *c*.1910: Long single-storey roughcast building with full-length glazed verandah facing the road, glazed with small-pane fixed lights over leaded windows, with elaborately carved spandrels to panels. At the core of this building is an old cottage that was owned by a Mrs Reid. It was remodelled in Tudor style by Sir Emerson Henderson of Sion Mills (where the family house was also Tudor), but about 1910 it was leased by William Johnston, "a short tubby man" who "always wore a pair of shorts and a Panama hat". Johnston already ran a restaurant in Belfast and the Garden of Eden Bus Co (which ran just two buses). He built a ballroom at the back and opened the *Cingalee Tea Gardens* for dancing, in a building which successfully managed to combine the black and white vernacular of Crawfordsburn with a fashionable whiff of the Raj. The next owner was Paddy Falloon, who named the rooms the *Chalet Cottage*, then made the tea rooms into a private club in 1937, in order to concentrate on the Inn which he had also acquired from Johnston.
See *Country Club, passim; Eakin; Spectator 7 Sep 1935.*

Nos.10-24: *c*.1830 and later: Irregular terrace of alternating single- and two-storey houses, most probably having started life as cottages that have been variously "riz up" and otherwise altered. Few have original features, the half-doors and external shutters being comparatively recent, but as a whole the terrace is charming with its unified paint scheme of white walls and black plinths and the subtle changes from one house to the next.

No.26: Crawfordsburn Orange Hall: 1882: Two-storey stucco building, three bays wide with additional bay added with scooped gable; panelled doors under entablatures, and windows mostly divided either horizontally or vertically. Originally a *Masonic Hall*.
See *Country Club p.26; Lawrence 9557.*

MAIN STREET, Groomsport

The main commercial street of Groomsport runs along the shore, numbered from The Point. At the turn of the century, much of it was single storey and still thatched, and dominated by the tower of the Presbyterian Church.
See *Lawrence 9555.*

No.1: *c*.1930: Roughcast bungalow with herringbone-timbered gables and verandah.

No.9: *c*.1850: Two-storey house with good Ards doorcase; originally stucco with quoins, now stripped and somewhat altered.

Nos.11-13: Moorings Court: 1990: Three-storey block of apartments with ornamental bargeboards relieving a very bland frontage. This replaced a pair of lined stucco HOUSES of about 1870 which had round-headed doorcases.

Beyond them stood a terrace of single-storey dwellings with small-pane sashes and roped thatch, probably cleared about 1930.
See *Lawrence 9555; Spectator 3 Jul 1986.*

Groomsport Presbyterian Church: 1843 and later: Gabled church with square-plan rusticated stone tower placed centrally in front, linked to the church at one side by staircase turret, and on the other by a two-storey extension. The tower has three stages, with a clock above a lancet window at the first stage and paired lancets at the second stage, all round-headed, and a parapet with finials and balustrade. The body of the church is smooth-rendered and four bays long, with a vestry to the back built in 1971. Groomsport's Presbyterian congregation was founded in 1841 and the Rev Isaac Mack was appointed its first minister. There was difficulty in obtaining land to build a church from either of the local landlords, and services were held in a house at No.17 Main Street until the widow of a Rev Andrews sold the church its present site and the church was built in 1843, with the tower added in 1863. A school was built behind the church in 1844-6, but demolished about 1950, and the Rev Mack was also responsible for building PROVIDENCE PLACE, four houses on the Main Street one of which served as his Manse (demolished 1962) and the Dufferin Villas in Bangor. An adjoining house to the east, VIRGINIA HOUSE, was purchased by the church in 1958 and promptly demolished, leaving the church rather gaunt and forbidding on its grass plot.
See *Lyttle pp.39-40; Nelson, passim; Spectator 28 Mar 1958.*

Nos.41-47: Spar, house, Post Office: *c.*1870: Row of two-storey stucco houses, now rather mutilated; the post office house retains a stone Ards doorcase and six-panel door.

No.49: The Anchorage: *c.*1890: Bold two-and-a-half-storey stucco house with ornamental bargeboards to dormers, pedimented first-floor windows - the central one having a small balcony - and paired round-headed windows at ground-floor level flanking the central door.
See *Crosbie p.40; WAG 3135.*

Nos.51-55: *c.*1890: Two-storey five-bay stucco building with stucco quoins, moulded doorcases with panelled doors, and shopfront complete with corbels and pilasters. Stone outbuildings at the rear.
See *Crosbie p.40; WAG 3135.*

No.57: *c.*1990: Plain two-storey rendered building with crowded windows. This was formerly GLASS'S HOUSE, a two-storey three-bay stucco house with crested ridge, modest pitched-roof dormers with trefoil-pierced bargeboards, and central Ards doorway with cobweb fanlight.

Schomberg Memorial: Headstone on stepped concrete plinth surrounded by granite bollards, to record the landing of "Field Marshall Frieveric Duke of Schomberg KG" who landed at Groomsport on 13 Aug 1689.

War Memorial: *c.*1925: White granite obelisk on stepped plinth.

Walter Nelson Hall: 1895, for Rev Latimer (of the Presbyterian Church),

later altered: Built by Mr Mawhinney of Newtownards for £485, this was a *National School* to replace an earlier building at the back of the Presbyterian Church; it was converted to a hall in 1962. Now a plain rendered building.
See *Crosbie p.40; Spectator 15 and 23 Nov 1962; WAG 3135.*

COTTAGES, probably 18th century in date, used to stand on the Main Street at this point before their demolition in the late 1960s. They were single-storey, whitewashed, and thatched or covered in tarred slates. Cockle Row, on the Harbour Road, still gives a flavour of what was not so very long ago a common type of house in Groomsport.
See *Crosbie pp.37, 40, 41; Lawrence 9555; Spectator 14 Mar 1958; WAG 1291, 2156, 3135.*

The Lock and Quay: 1906, later altered: Two-storey roughcast building with half-timbered gable, brickbonded dressings to openings and two gables to main street. Built by J McClenaghan to replace Robert Andrews' earlier thatched PUBLIC HOUSE on the site, it has been consideraby altered in recent years. Formerly the *Groomsport Inn.*
See *Crosbie pp.40, 41; Lawrence 9555; WAG 1291, 3135.*

Maxwell Hall: 1893: Small but rather grand hall of slightly-glazed red brick with terracotta eaves moulding to gable, intricately moulded tall brick chimney and bellcast roof to main hall behind. Built by Major Robert Perceval-Maxwell and given to the Parish Church in 1929 along with the adjacent Orange Hall which was demolished in 1934.

MANSELTON PARK
Cul-de-sac off Donaghadee Road developed about 1925 as *Manselton Drive.*

MANSE ROAD
A pleasant narrow hilly road developed in the late 19th century and deriving its name from the manse at its southern end; a Victorian cast iron obelisk pillar with dogtooth ornamentation standing nearby marks the end of the road which becomes a pedestrian lane through to Brunswick Road.
See *Lawrence 9554.*

Nos.23-29: *c.*1910: Group of two-storey roughcast houses with overhanging eaves; gable with duple windows at first floor over projecting hipped bay; two-storey porch with escutcheon; octagonal ground floor bay with turret roof. Some altered, but nos.23 and 29 intact, the latter with Arts and Crafts sashes.

No.2: *c.*1890: Two-storey double-fronted stucco house with two-storey bow windows and frilly bargeboard and eaves board.
See *Lawrence 9554.*

No.10: *c.*1900: Much altered two-storey double-fronted house with good bargeboard; nos.12 and 14 similar, the latter with tripartite windows; all with good stucco boundary walls and pillars.

Nos.16-18: *c.*1900: Terrace of two-storey stucco houses with two-storey canted bays, originally with horizontally divided windows in chamfered surrounds; nicely chiselled entablatures to doors. A detached house at no.20 is similar.

No.40: Simpson Family Resource Centre, formerly *The Manse*: *c.*1865: Two-and-a-half storey three bay house with central gablet; chimneys removed and windows replaced in plastic, leaving very little character, although the large grounds and trees are maintained. Built by John McMeekan and Mr McKimmons for Rev William Patteson.
See *Seyers pp.14-15.*

MARALIN AVENUE
Residential street laid out shortly before the war up the steep hill from Moira Park and downhill to Moira Drive. Most of the houses post-war, typically two-storey, rustic brick with hipped tile roofs.

Marina: See Piers.

MARINE GARDENS
On the inland side are the sweeping lawns in front of the even numbered houses in Princetown Road, followed by further terraces of grass, shrubs and tall pines, which accompany this splendid coastal walk in front of the rocks round Bangor Bay on land acquired by the council through Act of Parliament in 1905.

On the coastal side of the Esplanade, *Laird's* (originally Lenaghen's) *Boats* were a summer fixture that however involved no structure. Of another, the MARINE PAGODA near Pickie, we know little, but it was used by Miss Ella Rose's Golden Star Concert Party for their evening performances in 1904. Despite (or perhaps because of) her entertainments being "such as a brother can bring his sister, or a man his wife to, without fear of hearing anything vulgar", Miss Rose lost money and at the end of the season had to offer the exotic Pagoda for sale cheap, "suitable for storeroom or workshop, sound and portable." Various more permanent structures have however been developed over the years.
See *Crosbie pp.27-29, 35; Hogg 36, 87; Lawrence 2860, 3880, 11217, C6010; Milligan pp.10, 37; Spectator 17 Jun, 29 Jun, 12 Aug, 26 Aug, 14 Sep 1904, 16 June 1905; WAG 348, 2081, 3104; Welch 4, 17-19, 26, 31, 47; Wilson p.25.*

Pillars at the entrance: A pair of sandstone pillars was erected at the entrance to the Marine Gardens *c.*1915, to "give a finished appearance to the Gardens". Their presence was strongly resented by the argumentative James Thompson of nearby Martello.
See *Crosbie p.30; Spectator 9 June 1916; WAG 2701.*

Open-air Gospel Hall: A small arena with preaching-box just inside the gates to the Marine Gardens, where those "so dispoged" can savour a whiff

of brimstone with their ozone. Religious gatherings near Pickie were mentioned as early as 1905, and the stand is still there, run by the Bangor Congregational Church.
See *Crosbie pp.27-29; Spectator 7 Jul 1905; WAG 348, 2081, 3104.*

Pickie: Following the development of the Marina (see *Piers*) in the 1980s, which included the construction of a breakwater out from the Pickie rocks, a series of new tourist attractions have been developed including a lake for mechanical swans, a number of pavilions and a model railway. Unfortunately in the process much of the original seashore at this point has been covered over. However the rocky foreshore is still complete beyond the breakwater, including the dramatic *Man of War Gullet*, a steep rocky inlet marked by a rubble-stone wall along the path, and the *Amphitheatre* which was created near the old *Ladies Bathing Place*.

The Pickie Rock was a popular spot for bathing by the 1860s, when a bathing box was erected there every summer. In 1887 R E Ward erected a "small concrete and stone dressing house", probably similar to that remaining on Seacliff Road, and swimming contests were held from Pickie every year. Mixed bathing was permitted there in 1916, and in May 1931 the PICKIE POOL was opened with an "outstanding Aquatic Programme". The pond was 100 feet square, ranging from $2^1/_2$ to $7^1/_2$ feet deep, and had a 35-foot diving board. 1500 tons of rock were blasted out to form it, and 3000 tons of concrete used to build it. Pickie Pool was a fairly featureless single-storey building which had always struggled to make its chlorinated water even a few degrees warmer than the sea beyond. The march of time made heated swimming elsewhere economic in the 1970s, and Pickie was demolished about 1991. The present **fun park**, built in 1992-93, was designed by McAdam Design.
See *BT 12 Aug 1992; Crosbie pp.31, 32; Eakin; Hogg 78, 85; Lawrence 6198, 9540, 11215, 12231; Morton p.29; NDH 21 May 1887; Praeger p.71; Seyers p.7; Spectator 9 May, 23 May and 6 June 1931; Wilson pp.10, 28, 34.*

LADIES' BATHING PLACE: *c*.1905: Further along from Pickie at Skippingstone is what a recent publication rather charmingly described as "the Ruined Ladies' Bathing Place" on account of its condition; the accompanying buildings have now all gone, although part of the concrete enclosure remains at an inlet in the rocks.
See *Crosbie pp.34, 35; Eakin; Lawrence 4737, 11632, 11634 etc; Spectator 14 Sep 1906; WAG 347, 399; Welch 4, 49.*

Bandstand: *c*.1893: Moved from its former site in the Sunken Gardens when the McKee Clock was built, this octagonal bandstand with its cast iron railings, columns, spandrels and acroteria was set on a grassy mound near Skippingstone. It has recently been repaired. Nearby is a square-plan timber **Pavilion** with hipped roof that was part of the original layout of the *Amphitheatre* formed at this part of the Marine Esplanade. In the 1950s, archery butts were set up in the Amphitheatre. A little further round the coast is an inlet in the rocks where the waves suck in and out, licking at the entrance to *Jenny Watt's Cave*. Jenny's identity is uncertain: some say she was a smuggler who used the cave, others that she was a girl who ran up the cave to hide from soldiers at the time of the 1798 rebellion and was drowned when the tide came in. The cave was apparently intact till the men making the Marine Gardens "introduced" the stone at the entrance to a stick of dynamite, and it is now filled with sand and grit.

See *Crosbie pp.8, 25, 33, 34; Kirkpatrick pp.108-109; Lawrence 4721, 11216, 11632, 11635, 11991; Spectator 28 Sept 1906, 19 Mar 1915, 20 Aug 1927, 8 July 1950, 13 Feb 1986; WAG 343, 396, 3103, 3296; Welch 19, 49; Wilson pp.7, 21, 86.*

MARKETHOUSE SQUARE
Listed in the 1852 Belfast Directory, suggesting that at that time Main Street was more tightly developed and there was a "diamond" in front of the market house.

· MARKET STREET
Commercial street from Main Street to Castle Square, laid out as a lane to Castle Square and only developed itself during the interwar years. A car bomb in 1993 led to demolition and rebuilding of some comparatively old buildings.

MARYVILLE PARK
Street off Bryansburn Road laid out in the late 1930s with its back to the railway line.

MAXWELL PARK
A cul-de-sac laid out off Maxwell Road about 1925, with some pleasant houses from the 1930s.

No.3: *c.*1925: Two-storey half-timbered pebbledashed house; half-timbered gables set forward with ornamental bargeboards; rosemary-tiled roof.

No.8: *c.*1935: Two-storey roughcast house with half-timbered gable; irregular hipped roof; leaded lights.

MAXWELL ROAD
An extension of Princetown Road linking it to Bryansburn Road, which was built in the early years of this century, and developed increasingly densely as the years went on. Mature and well cared-for gardens.

No.1: Glendaro: *c.*1885: Two-storey double-fronted stucco house with iron balcony between two-storey canted bays, central pitched-roof dormer and ornamental bargeboard.

No.3: Innisvar: This was a two-storey double-fronted stucco house of *c.*1890, with two-storey bow windows and fretted bargeboard; demolished *c.*1990, it has been replaced by two new houses.

Nos.5-7: *c.*1890: Pair of three-storey stucco semi-villas with broad two-

MAXWELL ROAD

storey bow windows, half-hipped roof, and ornamental consoles to doorcases; doors mostly altered.

No.13: Downshire House: 1908, for John Thompson: Large two- and three-storey double-fronted stucco house surrounded by good mature gardens; iron balcony on timber kneelers over main door; hipped roof with ornamental clay ridge. Excellent grounds with mature trees. Possibly by Samuel Stevenson, who designed a pair of semi-detached houses on a nearby site for Mr Thompson in 1903.
See *App 458.*

No.19: *c.*1910, by W D R Taggart for S Kelly: Trim two-storey double-fronted red brick house with one bow window and one rectangular bay.
See *App 519.*

No.21: *c.*1910, probably by Ernest L Woods: Two-storey roughcast house with one bay single-storey and the other rising to a half-timbered gable.
See *App 468.*

No.23: *c.*1910, by E & J Byrne for C McBride: Severely altered 1997.

No.27: *c.*1910: Single-storey roughcast bungalow with rosemary-tiled roof, ornamental flourishes in gables and conical roof on corner turret. Heart-shaped leaded light in front door.

No.33: Haeremai: *c.*1925: Modest two-storey roughcast house made extraordinary by two-storey rectangular bays set at an angle to the building on each corner, with bulgy wrought iron balcony over trellised verandah below.

No.20: Braidwood: *c.*1935: Broad irregular two-storey house in rustic brick with artificial stone dressings, steel casements and some leaded lights.

No.30: The Manse: *c.*1908, by Ernest L Woods for First Bangor Presbyterian Church: Irregular two-and-a-half storey roughcast house with Arts and Crafts details including stained glass staircase window. The congregation's first manse had been beside the church in Main Street, subsequently converted to shops, and the second was at 4 Downshire Road, which had been built by Rev Patton at his own expense, so the church approved the £1300 expenditure on this building.
See *App 382; Spectator 21 Sep 1906.*

No.32: Glenview: 1912, by Ernest L Woods for himself: Irregular two-storey roughcast house with hipped and gabled rosemary-tiled roof and exposed rafter-ends; buttresses flanking front door. Woods was the Borough Surveyor, and responsible for a considerable number of Bangor's Edwardian buildings. Originally called *The Nook.*
See *App 638.*

No.38: Seafield House: *c.*1905, by George Browne of Browne Bros for George McCracken: Two-storey pebbledashed house (originally stucco) with two-storey bows to Bryansburn Road and single-storey canted bays to Maxwell Road. McCracken was a solicitor who married the Edwardian

13 Maxwell Road: a substantial Edwardian villa, built about 1908 away from the bustle of the seafront behind a screen of pine trees. Few of these buildings now survive unaltered. (Peter O Marlow).

32 Maxwell Road: Ernest Woods' own house of 1912, utilising the fashionable roughcast on a house that varies coped and eaved gables and has tall plain chimney stacks. (Peter O Marlow).

authoress LAM Priestley (see *8 Seacliff Road*).
See *App 235; Young Province p.292*.

MAY AVENUE
Narrow cranked road from Hamilton Road to Castle Street, developed about 1900 on the site of a former rugby pitch.
See *Seyers p.5*

Nos.11-29: *c*.1890: Two-storey two bay smooth-rendered houses with gently stepped roofs and strings. Some houses still have fielded doors and sash windows, and two houses have longer beautifully-kept front gardens.

Nos.12-20: *c*.1910: Terrace of two-storey rendered houses with continuous ground floor verandah and frilly eaves board; windows originally had six-pane upper sashes over plain lower ones, most now altered.

No.22: *c*.1915: The rump end of the terrace squeezed into a corner site with a truncated front elevation, but compensated by an ornamental first-floor panel of white pebbles and stained glass in japonaiserie margined panes.

MAZE PARK
Short street from Moira Drive to Maralin Avenue, laid out about 1939.

McKee Clock: See Quay Street.

McWILLIAM STREET
Street listed in 1852, probably off High Street.

MIDDLE ROAD   See *Central Avenue*.

MILL ROW
Sometimes known as *Mill Lane*, and formerly as *Burnside* on account of the stream running down it from a mill dam near present-day Springfield Avenue which powered a mill as early as 1625 (and centuries earlier had provided nocturnal ablutions for St Comgall). The stream was later culverted under the gasworks (see *Bingham Lane*) but, until its demolition to form a car park in 1983, there was at the Main Street end of the lane a three-storey gabled CORN MILL of rubble blackstone with red brick dressings in a complex that included houses for the kiln-man and miller. In 1914, the Council asked Robert Neill to fence off a derelict LIME KILN (of which they had formerly had a number) "for the prevention of accidents". There were "two small houses" in the Row in the 19th century, one being a blacksmith's shop. The upper end of the alley used to meet up with Bingham Lane; it is now diverted by the Flagship Centre, and returns to High Street.
See *BHS III p.38; Seyers p.4; Spectator 19 June 1914*.

MOIRA DRIVE
Developed from Hamilton Road to Broadway about 1935, and largely

developed before the war, on the edge of the former golf course. The ground falls sharply at the back of the houses on the west side, and many of the houses have rear basements fitting into the hillside. Although badly damaged during the war - the nearby white-painted clubhouse and golf course being mistaken by a German pilot for an airfield - the street has some pleasant thirties Stockbroker's Tudor at nos.34 and 38, and a number of white-rendered houses with half-hipped Westmoreland slate roofs, some with rather vogueish brick courses under the cills and Deco brick ziggurats round the front doors.

MOIRA PARK
Short street of mostly detached two-storey housing in roughcast and brick, from Moira Drive to Broadway, laid out and partly developed during the 1930s on the former golf course.

No.15: *c.*1935: White rendered house with stepped feature to door; leaded lights; steel windows to first floor; hipped Westmoreland slate roof.

MORNINGSIDE
Short road from Groomsport Road to the pedestrian portion of Ballyholme Esplanade, developed in the 1930s. Most of the houses are a standard design of two-storey semis in pebbledash on reddish cement, with two-storey canted outer bays terminating in half-timbered gables; cement trims with keystones to segmental-headed windows at ground and first floors; some Arts and Crafts windows survive, with leaded lights to fanlights.

No.22: Fairholme: *c.*1935: White painted smooth-rendered house with wrought iron balcony, red tiled roof and a striking large mushroom-shaped opening for French door facing the sea.

Mornington Park: See 97-113 Princetown Road.

Morrowvale Terrace: See 9-45 Holborn Avenue.

Mossvale: See 162-170 Seacliff Road.

Mount Oriel: See 31-37 Princetown Road.

MOUNT PLEASANT
Terrace below Princetown Road entered off from a lane between nos.40 and 42, with steep narrow railed gardens sloping down to the Marine Gardens. Along with the adjoining Mount Royal, this terrace set the tone for most of the later Victorian seafront and Princetown Road.
See *Crosbie pp.27, 28; Lawrence 2362, 3880, 11217, C6012; WAG 348, 3104; Welch 24, 27, 31.*

Nos.1-6: *c.*1875: Two-storey stucco houses, with two-storey canted bays. Nos.1-2, *Mooreleigh*, have simple doorcases and were raised an additional storey shortly after they were built; nos.3-6 have consoled doorcases, but are

spoilt by a modern box dormer.
See *Wilson pp.40, 85.*

Nos.7-10: *c.*1890: Three-storey terraced stucco houses with hipped three-storey canted bays. Sash windows in chamfered openings; consoled shallow pediments to first floor windows, elaborate consoled and dentilled doorcases; aedicules over first floor windows; quoinstones to no.10.
See *Lawrence C6010; WAG 348, 3104.*

MOUNT ROYAL

Terrace below Princetown Road entered off from a lane between nos.28 and 30, developed at the same time as *Mount Pleasant.*

Nos.1-6: *c.*1875: Terrace of three-storey stucco houses with vermiculated quoinstones, ornamental bargeboards and finials to the outer ones. Central pair of houses have large gables with moulded eaves and two-storey canted bays. Rather cramped dormers, an early move away from the skylight that would have illuminated most attics at this time.
See *Lawrence 11217; Wilson p.40.*

N

NEWTOWNARDS ROAD

The main road to Newtownards was in existence before the 19th century, but development on it was only sporadic till after the last war. Rising from the abbey as it does, the lower stretch was known as *Church Hill* around 1900. The road continues after the ring road.
See *Seyers p.47; Wilson p.50.*

No.62: The New Cemetery: The cemetery, which opened on 22 July 1899, has a good stone wall and sturdy iron gates and pillars. Half way down the cemetery is a small octagonal rustic brick **Gazebo** of *c.*1930, with weathercock on ball finial.

Rosemary Street Baptist Church: 1963-64: Rustic brick hall with copper roof and fleche.
See *Spectator 18 Sep 1964.*

No.76a: This was the location of THE INKPOT, a whitewashed two-storey stone house with two lancet doors and a small window over the door in an octagonal end, with a single chimney on the ridge. It was considered to be a toll house, and when it was made the subject of a closing order, the possibility of protecting it under Ancient Monuments legislation was explored. Sadly, because there was "no positive evidence" that it was a tollhouse, it was demolished in 1960.
See *Spectator 26 Sep 1958, 15 Apr 1960.*

Bangor Fire Station: *c.*1990, by Gifford & Cairns: Beyond the Circular Road. The main block of the fire station is symmetrically built of pinkish brown brick with a dark-glazed clerestorey and oversailing eaves; three large bright red fire-engine doors are flanked by full-height glazed bays. Alongside stands the traditional tower.
See *Perspective Sep 1992 p.3.*

Lead Mines: See Main Street, Conlig.

O

OAKWOOD AVENUE
Short road from Belfast Road to Beechwood Avenue, laid out about 1925. Nos.2 and 8, built in the late 1920s, were quite stylish hipped-roof bungalows, but now have plastic windows.

OLD BANGOR ROAD, Conlig
The road between Bangor and Newtownards now by-passes the village of Conlig, but portions of the Old Bangor Road survive at each end of the village.

OLD BELFAST ROAD
From Springhill, the original Belfast Road was superseded by the present Belfast Road with the completion of the ring road about 1975. The old road runs in parallel with it to Clandeboye.

Nos.214-216: Ballyvernott Cottage: Single-storey cottages which appear on the 1833 OS Map, but are much changed.

Viscount of Clandeboye: *c.*1975, by H A Patton: Octagonal public house with a butterfly roof forming an "eyecatcher" on the hill beside the main road. Entered across a bridge at first floor level from the shopping centre, each face of the octagon at first floor level is multiplied twofold.

Springhill Shopping Centre: *c.*1970, by H A Patton: Early out-of-town shopping centre with corrugated metal roof, taking its name from the farm formerly on site, Spring Hill (see *Springhill Road*).

Olga Mount: See 55-59 Bryansburn Road.

OSBORNE DRIVE
A street of mostly Edwardian and inter-war houses of pleasant character developed about 1910 from Bryansburn Road to Brunswick Road.

No.5: White Lodge: *c.*1925: Two-storey white roughcast house with stained glass upper lights to casement windows; projecting rafters and roughcast wallhead chimneys.

OSBORNE DRIVE

No.15: *c.*1925: Two-storey red brick house with white roughcast first floor. Palladian window in centre of front elevation (presumably at staircase), with dentilled hood moulding over. Many windows unfortunately now altered.

Nos.2-12: *c.*1910: Group of two-storey double-fronted red brick detached houses with roughcast first floor, canted ground floor bays, block-bonded quoins and curious partially glazed doors. All now much altered.

No.14: *c.*1925: Two-storey house with canted bay extended to form a verandah, with trellised spandrels; segmental-headed tripartite windows with tinted glass small-pane upper lights to casements; projecting rafter ends.

Nos.18-20: *c.*1925: Double-fronted semi-bungalows with red rosemary-tiled roof with terracotta finials to hipped bays, and steep half-timbered gables over recessed doors.

OSBORNE PARK
L-shaped avenue from Osborne Drive to Brunswick Road, mostly of two-storey semis, laid out about 1925 and built up before the war.

Nos.32-34: *c.*1930: Pair of hipped two-storey semis with stucco ground floor and roughcast first floor; tripartite and duple windows; stained glass toplights to square ground floor bays under projecting porch roof.

P

THE PARADE: See *Quay Street*.

PARK AVENUE
Short road from May Avenue to Park Drive, laid out about 1910.

Nos.1-17: *c.*1910: Two-storey smooth-rendered terrace with continuous ground floor verandah joining bay windows and forming porches above doors. Doors have ogee heads, and cartouches in plaster.

PARK DRIVE
Slightly angled single-sided street from Hamilton Road to Castle Street, laid out about 1920 and looking through a fringe of trees into Ward Park.

No.2a: Church of the Nazarene: *c.*1980: Office building converted to church use in the early 1980s, and described as "a holiness church in the Wesleyan tradition."
See *Spectator 19 Mar 1998*.

Nos.6-16: *c.*1925: A curious survival of Edwardiana, stucco semis with first floor verandahs with windbreaks. Nos.14 and 16 have elaborate cast iron verandahs with fretted spandrels. Roofs descend over balconies in a bellcast.

14-16 Park Drive: post-war houses with first floor verandahs under bellcast roofs looking on to Ward Park through lacy Edwardian balconies. (Peter O Marlow).

22 May Avenue: an ornamental cottage squeezed on to a corner site but managing to make the most of it. (Peter O Marlow).

25 Primacy Road: an unusual gateway manufactured from spinning wheels and mangles.

PARKMOUNT

Cul-de-sac off Park Drive, laid out about 1928 and developed by 1935, taking advantage of an elevated site to look on to Ward Park.

No.6: *c*.1930: Detached roughcast bungalow, with hipped red-slated roof and stained glass toplights.

Patteson Terrace: See 2-12 Croft Street.

PICKIE: see *Marine Gardens.*

PICKIE TERRACE

Like the earlier Mount Royal and Mount Pleasant, this is a terrace accessed from Princetown Road (at nos.70-72) but looking over Bangor Bay.

Nos.1-6: *c*.1885: Three-storey terrace of six stucco houses, dominated by the four-storey square-plan Italianate tower of no.1; three-storey canted bays to most houses with gables above nos.1 and 6. No.2 had a charmingly bulgy lancet arch at its entrance gateway from the sea, which seems to have been demolished along with the rest of the seaside boundary during the improvements to the Marine Gardens. Nos.1-3, which have fretted heads to opes and some margin-paned windows, were built as the *Pickie Rock Hotel*, but became the *Bangor Collegiate School* about 1925. Nos.4-6 are relatively plain but retain fielded doors.
See *Crosbie p.29; Eakin; Lawrence 9538; Seyers p.14; WAG 2081.*

PIERS

178 In 1620 King James had granted Sir James Hamilton the right to establish a maritime port in Bangor, and the Custom House of 1637 was evidence of Hamilton's will to pursue his opportunities. However, as early as 1664 Bangor no longer appeared on the table of Irish Customs and Excise Returns and, possibly due to mismanagement by Hamilton's heirs, the port had obviously declined. About 1760 a small pier was built with the aid of a grant of £300 provided by the Irish Parliament in 1757, and in 1791 the Collector of Customs at Donaghadee stationed a clerk at Bangor to deal with the growing trade. Nimmo in 1822 describes Bangor as being much frequented by trade with Scotland, notably fishing vessels and the shipping of live cattle. By 1830, coal was being imported in some quantity, and the coastguard station at Bangor had five men and an officer, with three other stations along the eight-mile coast of Bangor Parish.

Nevertheless the harbour was described in the 1833 OS Memoirs as "a very bad one. At low water it is left completely dry. There is one pilot in the town who is connected with an assurance company, and whose principal business it is to put safe out of Bangor Bay the vessels which have put into the harbour. For this he receives 2s 6d". In 1844 "fifteen sail of carrying vessels, three stout fishing wherries, and a number of yawls" used Bangor.

PIERS

In 1865 the contractor who had just finished constructing the railway to Bangor extended the narrow old pier to assist in the increased traffic, which by now included imports of limestone, timber, coal and iron and exports of cattle, lead and copper ores, tiles and bricks. The two chief companies operating from Bangor in the late 1860s were Robert Neill's cargo boats and J & R Brown's passenger paddle steamers running thrice daily to Belfast. Up till the Great War, many commuters to Belfast preferred to use the steamers rather than the train in the summer, and their railway tickets were often interchangable with steamer tickets. However, in 1885 even W G Lyttle, giving the town a hard sell in *The Bangor Season*, was forced to admit that "the pier... could be greatly improved, the present structure being... under certain conditions, absolutely unapproachable and dangerous." By 1892 Robert Ward was requesting Lord Dufferin's assistance in obtaining a Treasury loan to "erect a new pier here for steam passenger traffick & also merchant vessels". Mr Macassey had designed a structure in concrete to low water mark, with the remainder of the total 800 foot length in "green hart piles" since concrete would be too expensive for the whole length. That a new pier was necessary is evident since Ward quotes a figure of 600,000 passengers landing at Bangor in 1890-91. In 1896, the new pier was duly built at a cost of £24,000 and "furnished with a band stand". Thus by 1900, Bangor had three piers in its Bay - the South Pier or *Neill's Pier* (the Coal Depot) off Quay Street near the Bank, the *Old Pier* at the Customs House, and the *New Pier* just to the North.

At the time of the construction of the New Pier, the old mills were cleared and "a handsome esplanade" was created at the bottom of High Street, known as the *Sunken Gardens*. Surrounded by elegant railings and containing the Coates Memorial, the bandstand, and a rather ornamental Coal Office, the Esplanade (see *Quay Street*) was a source of considerable local pride. Not for long though - in 1914 the former town clerk, James McKee, left money to build a clock, and the Co Down Spectator reported that the "eyesore" bandstand would be removed to make way for the clock. The Coal Depot, which was a far greater disfigurement, was acquired from Robert Neill by the Council in 1931, cleared of coal sheds and renamed the *South Pier*; and the central pier was extensively repaired in 1935.

In April 1914, the pier was the scene of a night of dramatic activity when eighty tons of rifles were landed in secrecy from a mist-shrouded *Mountjoy*. The town was sealed off at midnight by around a thousand Volunteers opposed to the threat of Home Rule, and shortly after over a hundred cars were lined up along the Esplanade. "Is Bangor going to be taken?" asked an anxious visitor. "Bangor's tuk" replied a laconic volunteer. The arms were transferred to the vehicles for distribution around the countryside (a quantity went to an underground tunnel at Seacourt in Princetown Road), and the town had reverted to normal by 6am. The following Sunday, the concerned rector preached in St Comgall's about the evils of civil unrest, unaware that on the sounding board above his head reclined a portion of the *Mountjoy*'s cargo.

PIERS

In 1981-83 substantial changes were made to the harbour with the building of the £2m *North Breakwater* designed by Kirk McClure & Morton, which won a Concrete Society award. The *Marina* itself was developed from about 1985 till the early 1990s, and includes an area of ground reclaimed from Bangor Bay to provide landscaped car parking. It has also absorbed the former Sunken Gardens on the seaward side of Quay Street, including the McKee Clock and Coates Memorial. The Marina was completed with an additional breakwater from the rocks near Pickie and an extension to the Central Pier to enclose floating pontoons and service small craft. This provides a sheltered anchorage for boats, and fishing platforms for promenaders - provision was even made for the rare black guillemots which always nested in the pier - although the vista of concrete cubes from Seacliff Road does not have the charm of the old structures, and the adjacent car-park has had the unfortunate effect of leaving the Custom House and old Harbourmaster's house stranded high and dry.

The marina is a major part of North Down Borough Council's seafront development scheme designed to inject finance into the presently very run-down Queen's Parade area, and while aspects of it have been very successful, the inevitable delays in completing major redevelopment schemes has left many of the buildings on the seafront without investment in the medium term. While it is apparent, too, that local yachtsmen are using the marina - Ballyholme Bay, which used to be thick with boats in the summer, is now almost empty - the amount of tourism generated by the development must surely be modest, and there is inevitably unsightly security fencing around the pontoons. The "Scotchies" who used to come over during Glasgow Fair Week every year now go elsewhere, and are unlikely to be attracted back by the large area of immobile boat-parking. One of Bangor's most individual tourist assets has always been its interesting and readily accessible seaside, particularly the superb rocky foreshore along from the Kinnegar to Stricklands, and this unspoilt natural beauty must not be lost to commercialisation. It has to be said that the new planting and fountains are thriving, and that the car parks and parked boats seem to attract as many sightseers as the old bay did with its piers, rowing boats and rocks. More controversially, however, about 1992 the new *Bregenz House* was built well out into the bay on reclaimed land, further blocking the view of the sea that used to be enjoyed from both Main and High Streets: "I couldn't see the sea at all when I reached the bottom of Main Street" is now a common comment.

See *Arch Surv p.395; BT 25 May 1989; Crosbie pp.31; Eakin; Green p.75; Hogg 78; Lawrence 4739, 6106, 6198, C2359, C6017, I557, I2683, I3594; Lyttle p.33; McCutcheon pp.148-49; OS Mems p.25; PRONI D.1071; SBP 17 Oct 1991; Seyers pp.8, 11; Specify Jun 1983 pp.35-37; Spectator 1 May 1914, 1 Aug 1931, 14 Jan 1933, 5 May 1988; WAG 399; Welch 1, 2; Wilson pp.8, 10, 11, 22, 26, 27, 28, 30, 33, 36, 40, 49, 86, 87.*

Bregenz House: *c.*1992, by McAdam Design: Three-storey building in pinkish-brown ribbed reconstituted stone presenting six dormers and a

Diocletian window to the town.

PITCAIRN AVENUE
Cul-de-sac off Bloomfield Road, developed during the 1930s.

POPE'S LANE: See *Chippendale Avenue*.

PRIMACY ROAD
This road is a loop off Bloomfield Road towards the Gransha Road. From before 1830 until 1980, the irregular row of low single-storey cottages which made up *The Primacy* had the character of a small hamlet independent of Bangor and looking to the open country, but the suburban sprawl has now engulfed it and the cottages look on to a neat row of red brick spec houses instead of countryside.

No.3: White Lodge: *c.*1910: Roughcast house with stone gate pillars, set behind a good screen of trees. This was probably the *School House* for the hamlet of Primacy which closed in 1960. Until its construction, one of the adjacent cottages seems to have served as the school.
See *Spectator 8 Jul 1960.*

Nos.5-27: pre 1833: Although none of these cottages is in anything like its original condition, they have considerable antiquity and no little charm as a group. No.25 has a wonderful gate (made about 1985) with spinning wheels and iron mangle wheels woven into a trellis, while no.27, rather surrealistically given its distance from the sea, has a boatyard in its side garden.
See *BHS I p.20.*

PRIMROSE AVENUE
Road from Central Avenue to Primrose Street and Gray's Hill, laid out about 1880 but little developed till after the first war.

No.1: Spire View: *c.*1905, for H Sloan: Double-fronted stucco house with canted ground floor bays and steep gablets with projecting eaves and decorated apex board over twin round-headed windows; frilly eaves board between gablets.
See *App 445.*

Nos.11-13: *c.*1905, for William Hanna: Pair of two-storey semi-detached houses with hefty verandahs, porthole windows and roof gables.
See *App 396.*

PRIMROSE STREET
Street running uphill from Southwell Road to Dufferin Avenue parallel with Gray's Hill laid out by 1903 (on a former potato field but perhaps one with primroses round the sides) but mostly developed during the 1930s.
See *Seyers p.2.*

Nos.23-29: *c.*1925: Terrace of two-storey rendered houses with porthole windows at first floor; double sash windows to first floor, tripartite to ground floor.

Nos.10-12: *c*.1890: Pair of two-storey semi-detached stucco houses with linked central doorcases.

PRINCETOWN AVENUE
Steep road winding up from Somerset Avenue to Princetown Road, laid out about 1905.

Nos.1-11: *c*.1910: Three pairs of two-and-a-half and three-storey semi-detached houses in red brick with red sandstone lintels and cills, pebbledashed first floor, sturdy dormers and chimneys. Brick chimneys rise from the roof slope rather than ridge. Until recently the woodwork was uniformly painted dark green.

PRINCETOWN ROAD
Containing the most consistently fine of Bangor's villas, this road was developed in the last decades of the 19th century along a gently curving line following the shape of the Wilson's Point peninsula between Bangor Bay and Smelt Mill Bay. The mature gardens, many with good trees, complement the fine stucco houses, and at the Bangor end there is a row of pollarded limes and horse chestnuts along one side of the street.
See *Seyers pp.13-14; Lawrence 9547, 9548, C6010*.

Nos.1-9: *c*.1875-80: Terrace of two-storey stucco houses with two-storey canted bays, set back from the road and at an angle to it. Originally with dormers, horizontally-divided sash windows and panelled doors, though some details have been lost. Nos.7-9 are slightly different, with porticos and quoins. An early photograph shows nos.1-5 (*Somerset Terrace*) complete but no other development along Princetown Road; however nos.7-9 must have followed shortly after.
See *Lawrence C6015; Wilson p.40*.

Nos.11-37: *c*.1885: Three terraces of two- or three-storey stucco houses, with two-storey bays, bowed at nos.11-21 but canted at nos.23-29 and 31-35. Some houses still have original sash windows and panelled doors. No.31, which has been pebbledashed, retains bargeboard and finial. No.37, a taller house with bow windows, completes the terrace. Stone boundary wall to road, and mature trees. Nos.31-37 were known as *Mount Oriel*, the rest as *Shandon Terrace*.
See *Lawrence C6015*.

157 **Nos.39-41: Medora:** *c*.1895: Pair of three-storey stucco semi-villas with slightly Art Nouveau detailing and irregular design including unmatched gables; timber verandahs; a flowing centre feature in the stucco encloses triple ground floor windows, double first floor and round-headed second floor windows.

39-41 Princetown Road: an unusually irregular pair of houses for their date of about 1895, with flowing Art Nouveau elements on the facade.
(Peter O Marlow).

75-77 Princetown Road: dating from c.1890, one of the best-preserved pairs of semi-villas on the road, with original fenestration and a very grand roofline of barrel-roofed dormers and substantial chimney stacks. (Peter O Marlow).

30-40 Princetown Road: this pretty terrace of about 1880, with its tight rows of finialed dormers and ground floor bay windows, is on a more modest scale than most of the Princetown Road. (Peter O Marlow).

Little Clandeboye, Main Street, Conlig: this fine neo-Tudor mansion of about 1820 became a mental asylum in the 1930s and later was an imposing ruin for many years before being demolished. (Hogg Collection).

Nos.47-61: Mayfield: *c.*1893: Terrace of three-storey stucco houses with two-storey canted bays and mutual shouldered gables originally ornamented with ball finials and urns, elegantly set back from road with a common lawn and hedge.

Nos.63-65: Thornbrook: *c.*1890: Pair of three-storey stucco semi-villas with two-storey canted bays topped by fretwork and ball finials; stucco shield over door. Set at the back of a high lawn overlooking the road.

Nos.67-69: Arumah: *c.*1890, altered by McAlister Armstrong Partnership 1985: Two-storey stucco double-fronted house formerly with pointed windows in gables; now largely rebuilt and extended.
See *UA Dec 1985.*

Nos.71-73: Florenceville: *c.*1895: Two-storey stucco semi-villas each three bays wide, with steep gables.

Nos.75-77: Clare and Cooldara: *c.*1890: Two-and-a-half storey stucco semi-villas with very grand rooflines involving massive fluted and corbelled stucco chimneys, barrel-roofed dormers, balustraded tops to porches, and canted bays. Original sash windows. No.77 was originally called *Closeburn*.

Nos.79-81: Bayswater: 1900, by S P Close: Two-storey semi-villas with two-storey canted bays at each end; central windows of bays bipartite and main windows tripartite with slender pilaster colonettes on the mullions; Arts and Crafts sashes.
See *App 6.*

No.83: Avoca House: 1998, for Higginson Homes: Three-storey block containing nine apartments, with four similar apartments (*Farnham Lodge*) to the rear of the site. Smooth-rendered buildings with plastic windows, some balconies. This replaced the original AVOCA of *c.*1890, which was a two-storey double-fronted stucco house with two-storey canted bays; it had keystones to all windows (the sash windows of which had been replaced with aluminium) and an ornamental bargeboard to gable. It lay derelict after some years in educational use.

Nos.85-87: Glenoe: *c.*1892: Two-storey stucco semi-villas with steep gable roofs over two-storey canted bays.

Nos.89-91: Ardlussa: *c.*1895: Two-storey stucco semi-villas with two-storey bow windows rising to conical roofs with cast iron finials; tripartite ground floor window; windows partly altered.

Nos.93-95: Rostellan: *c.*1910: Two-and-a-half storey stucco semi-villas with two-storey canted outer bays and central three-storey gables; crested terracotta ridge.

Nos.97-113: Mornington Park: Screened from the road by an enormous fuschia hedge and with its own internal roads entered through stucco piers at each end, this group of houses dates between 1890 and about 1910. Nos.97-101 is a terrace of two-storey stucco houses with canted bays dating from

PRINCETOWN ROAD

about 1905; no.103 and the pairs of semi-villas at nos.105-111 are three-storey with full-height canted bays, built about 1895. No.3 Raglan Road is contemporary and part of the same garden suburb. STONELEIGH, at no.113, was an Arts and Crafts house built about 1910, with large half-timbered gable containing a shallow bow oriel at the second floor, and had casement windows with stained glass toplights. It was demolished *c*.1985 to make way for two smaller houses.

Nos.10-16: Clonallon: *c*.1900: Terrace of two-and-a-half storey red brick houses with mutual half-timbered and roughcast gables oversailing two-storey canted bays; small-pane upper sashes; quarry tile paths.

Nos.18-24: Westwood: *c*.1900: Terrace of two-and-a-half storey red brick houses with two-storey canted bays, brick dormers and terracotta panels.

Princetown Terrace: see separate entry.

Nos.26-28: Possibly *c*.1860, much altered: Pair of low roughcast semis with basements. Cast iron railings survive at steps.

Mount Royal: see separate entry.

Nos.30-40: Ardmore Cottages: *c*.1880: Terrace of one-and-a-half storey stucco houses with frilly bargeboards to deep-eaved dormers over ground-floor canted bays; corbelled flat entablatures over doors; all windows and doors unfortunately altered but well worth restoring.
See *Lawrence 9548*.

Mount Pleasant: see separate entry.

Nos.42-48: *c*.1880: Two pairs of low one-and-a-half and two-storey houses in a terrace, both much altered; with gablets over round-headed windows at first floor of nos.42-44.
See *Lawrence 9547, 9548*.

Nos.50-54: Ardeen: *c*.1905: Terrace of two-storey hipped-roof stucco houses with two-storey canted bays; fluted pilasters to doorcases.
See *Lawrence 9547*.

Nos.56-58: Martello: *c*.1885: Two-storey semi-villa presenting its main elevation to the sea, with hipped wings set forward and fronted with two-storey bow windows. In 1905 a servant girl "jumped through the window of a second storey room" of the house and fell into the yard; the unfortunate girl suffered no injury but "was certified to be a dangerous lunatic" on account of a religious mania, and consigned to the Down District Lunatic Asylum. This was later the home of James, one of the two argumentative Thompson brothers, who took his dispute with the Council over the Marine Gardens to the House of Lords. Former **stables** to no.56 front the Princetown Road, constructed in random rubble with red brick dressings and frilly bargeboard; good random stone wall alongside no.56 alongside the Tennyson Avenue route to seafront.
See *Crosbie pp.27-29; Hogg 87; Lawrence 2362, 2860, C6010, C6012; Milligan p.10; Pike p.302; Spectator 17 Feb 1905; WAG 348, 2081, 3104; Welch 24, 31; Wilson p.69*.

PRINCETOWN ROAD

Nos.60-62: Augustaville: *c.*1887: A magnificent pair of stucco semi-villas, possibly the finest on the Marine Esplanade - for, like its neighbours, Augustaville presents its best face to the sea; three-storey with gabled windows below frilly bargeboards and finials; two-storey bow windows at front and sides; mutual Doric portico and rusticated ground floor; quoinstones ranging from the conventional at second floor level through a pilaster form at first floor to vermiculated quoins at ground floor; stucco chimneys with splendid tulip pots; stone boundary wall with bold balustrading and ball finials to pillars. Built by Robert Russell, secretary of the gas company, whose family ran a drapery business in the town. According to Seyers, the development company set up to build Princetown started to build a hotel near here, and "had a tool house built where Augustaville is" before they went bankrupt, after which a Mrs McAlpin had tea rooms on the site.
See *Crosbie pp.27-29; Lawrence 2362, 2860, C6010, C6012; photograph in Bangor Heritage Centre dated 1887 showing Augustaville under construction; Seyers pp.3, 13; Spectator 5 Sep 1931; WAG 348, 2081, 3104; Welch 31.*

No.64: Princetown House: *c.*1890: Two-and-a-half storey stucco double-fronted house facing the sea, with two-storey canted bays, vermiculated quoins, Tuscan porch, sheeted eaves on decorative corbels, stucco chimney-stacks; windows unfortunately modernised.

No.66: Princetown Villa: dated 1900; by J C McCandliss for James Campbell: Two-and a-half storey stucco double-fronted house facing sea with two-storey canted bays with dentilled and balustraded tops; central stucco panel above first floor level with design of ivy leaves; elaborate cornice with brackets alternating with stars; triple layered bargeboards topped by terracotta finials to dormers; stone boundary wall. Dated in quatrefoil at rear.
See *App 3.*

No.68: Princeton: 1900, by J C McCandliss: Two-storey double-fronted red brick villa with tripartite windows with colonettes on the mullions, above ground floor canted bays.
See *App 2.*

No.70: *c.*1995: Three-storey rendered apartment block with balconies, bow and bay windows facing the sea. While the general scale and massing of the building fits with its neighbours, it lacks the vigorous detail of its Victorian predecessor. This site was formerly occupied by ROCKVILLE, a two-and-a-half storey double-fronted stucco house of about 1880, with two-storey canted bays flanking a central Tuscan porch; dentilled string courses; quoinstones vermiculated at ground floor level; and frilly bargeboards framing round-headed dormer windows. Fortunately its stone boundary wall towards the sea, with remarkable corner pillars topped by obelisks supporting terracotta urns, has been retained.
See *Eakin; Lawrence 2864.*

Pickie Terrace: see separate entry.

PRINCETOWN ROAD

Nos.72-74: *c*.1890: Pair of two-storey stucco semi-detached houses with ground floor canted bays and round-headed windows to first floor.

Nos.76-82: Granville Terrace: *c*.1895: Terrace of four two-storey stucco houses with two-storey canted bays.

Lorelei: see separate entry.

Nos.92-94: Claremont: *c*.1880: Pair of two-storey stucco semi-detached houses with ground floor canted bays, round-headed windows to first floor and bracketed cornice; windows and doors all altered.

Nos.96-98: Kensington Villa: *c*.1880, built by Sandy McFerran for John Neill: Two-and-a-half storey pair of stucco semi-villas overlooking the Bay with two-storey bow windows and mutual rectangular bay; no.98 somewhat altered. Fine stone boundary wall on all sides, with outbuildings to Princetown Road and corner obelisks on the sea side.
See *Seyers p.35.*

Nos.100-102: Glandwyr: *c*.1900: Two-and-a-half storey stucco semi-villas facing the sea with ornamental bargeboards and barrel-roof dormers; windows unfortunately modernised.

Nos.104-106: Minavon and **Ellenville:** *c*.1880: Two-and-a-half storey stucco semi-villas facing the sea with broadly battered stucco chimneys on crosswalls, canted two-storey bays, quoinstones. In 1926, the young Rev W P Nicholson, who lived at Ellenville, took a successful libel case against the *Sunday Express* for its headline "Parson with Bare Knees. Object lesson to Flappers. Naked Chest", alleging he had bravely appeared topless at a gospel meeting in Portadown to show young girls how they looked to others when they wore low necks and sleeveless frocks.
See *Lawrence 11632; Spectator 30 Jan 1926; Welch 2.*

No.108: Princetown Lodge: *c*.1906 by Ernest L Woods: Two-storey house with half-hipped roof; considerably altered. Built for R J Woods, damask designer, opera singer and painter, who in 1952 still wore a bow tie and had "an eye-glass fixed firmly in his right eye".
See *IB 10 Mar 1906 p.198; Pike p.180; Spectator 28 Sep 1906, 24 May 1952, 3 Aug 1957; Young p.227-28.*

No.114: Innisfail: *c*.1906, but considerably altered: Substantial two-and-a-half storey roughcast house with corbelled chimneys.
See *Crosbie pp.33, 34; Lawrence 11635; WAG 343, 396; Welch 49.*

No.118: Glenbank: *c*.1890 for the Connor family, Belfast merchants: Two-storey red brick house of irregular plan, with stone dressings and quoins; handsome fretted gable boards over Ionic porch and subsidiary gablets; consoles to window cills; windows a mixture of plain and round-headed sashes; roof partly hipped. At the bottom of the garden, **Glenbank Tower** is an octagonal turret (with extension) in rendered brick, while outbuildings to Seacourt Lane share the fretted gable-boards of the main house. Former

Augustaville, 60-62 Princetown Road: probably the finest pair of semi-villas in Bangor, occupying a very prominent site overlooking Bangor Bay. Dating from 1887, with crisp stuccowork. (Peter O Marlow).

Two Victorian urns: on the left, the terracotta urns on boundary obelisks from the now demolished 70 Princetown Road; on the right, the stone gate pillars of Seacourt, 120 Princetown Road. (Peter O Marlow).

118 Princetown Road: although Glenbank was built of brick rather than stone, it was nearly as grand as the neighbouring Seacourt, though built later in about 1890. It is now in flats. (Peter O Marlow).

120 Princetown Road: built about 1865, Seacourt was the finest house in Bangor until its subdivision into flats and the development of its gardens in the 1990s. The elegant bow window looks over Bangor Bay. (Peter O Marlow).

octagonal brick **gate lodge** greatly altered. Tennis courts remain, though the house has been divided into flats, and the grounds developed with bungalows in 1987. Important rendered wall nearly twenty feet in height to the south forms an enclosure to the Marine Gardens, while the side wall to Seacourt Lane is some twelve feet high in rubble stone, and to Princetown Road stone gate pillars and handsome timber gates complete the picture.

See *Crosbie pp.33, 34; Dean p.77; Lawrence 9539, 11216, 11635; PRONI D.1898/1/5; Spectator 16 April 1987; WAG 343, 396; Welch 19. 49; Wilson p.86.*

No.120: Seacourt: *c*.1865: The grandest house on Princetown Road, occupying what was once a large and prime site on Wilson's Point. The house itself is two storeys in height, of stucco with stone decorations and cills. The main front is three bay, with a rather cluttered channelled ground floor featuring well-spaced tripartite windows flanking the central Doric porch whose fluted columns support a small pediment; a fretwork stone balcony set out on hefty stone kneelers forms the cornice above the ground floor; first-floor windows inset to wall in slightly Egyptianesque surrounds; deep cornice with consoles supporting low parapet, decorated by rows of acroteria. The garden front overlooking the sea is also symmetrical, with a central two-storey bow window and boldly projecting balcony as on main front. *163, 164*

Despite recent alterations, the richly carved timber "pulpit" at the foot of the staircase (inserted *c*.1900), magnificent panelled doors, spiky vegetable plasterwork and marble fireplaces apparently remain intact. The west front has a fine chunky timber conservatory with fluted columns to its porticoed entrance and stained glass depicting birds in the roof lantern. The rear of the house is three-storey, but similar in height to the main building. Handsome stone **gate pillars** ornamented with acroteria and square draped urns (and at present a little barbed wire) mark the curved entrance sweep; there is a hipped-roofed **gate lodge** of *c*.1920, formerly the gardener's cottage but currently derelict. The grounds have largely been built on, but originally included a castellated stone wall with corbelled turrets at intervals; castellated walled garden; small formal garden with box hedging, palms and tree heathers; fernery, rose garden and a "ruined castle". **The Tower**, set in a bank of tall windswept trees at the bottom of Brompton Road, was built at the turn of the century to watch the yacht racing.

Seacourt was built about 1865 for Foster Connor, a Belfast linen merchant whose initials are still etched on the front door of the house. The architect is unknown, though Dean suggests that the gate pillars are in the Greek Revival style favoured by the Glasgow architect James Hamilton, who was certainly working for the linen merchants Ewarts in Belfast in 1869. In 1895 Samuel Cleland Davidson, the founder of the Sirocco Engineering Works, bought the house and 18-acre grounds for £5000. On his death, the house passed to his younger daughter Mrs Haddow, and in 1972 it was purchased by Down County Education Committee to become a Teachers' Centre. Sadly since 1989 when the house and its 6.5 acres were sold by the Education and Library Board the house has been divided into flats, and the grounds have been sold off and developed piecemeal.

Samuel Davidson was a keen amateur photographer, and also played the violin and piano. He had gone to work at a tea plantation in India at the age of 17, teaching himself Hindustani and navigation on the way out. Conditions there were primitive, and two of the three companions who went out with him were dead within five years. While in his twenties, he patented the first mechanical tea drying process, and went on to develop innovative mechanical fans and other tea equipment, all of it manufactured in Belfast but soon sold across the world. During the First World War, Sirocco fans were used in British, German and American naval warships, as well as in the London Underground. Davidson was knighted shortly before his death in 1921, by which time he had over 120 patents to his name. With his mechanical bent, he introduced to Seacourt a kerosene pump (now in the Ulster Museum) to take water from a well to a tank in the roof, simultaneously operating a butter churn. He was a keen yachtsman, and entertained, among other distinguished guests, Sir Thomas Lipton, the challenger of the America's Cup. Lipton was wealthy but the traces of his upbringing as a Glasgow barrow boy sometimes showed through. The last private owner of Seacourt remembers Lipton complimenting her grandmother on a meal and saying "Mrs Davidson, *them peas was lovely*".

See *BT 18 Jan 1989; Dean p.93; Mons Record; Pike p.308; Seyers p.14; Spectator 8 May 1986; Young p.216.*

PRINCETOWN TERRACE

Cul-de-sac entered from a lane off Princetown Road, developed about 1900 with a terrace facing the sea.

Nos.1-2: *c.*1900: A pair of three-storey stucco houses with bow windows; ornamental bargeboards to finials.

No.3: *c.*1900: Narrow three-storey brick building. **Nos.4 and 5** are drastically altered.

No.6: 1932, by Thomas Callender for Hotel Pickie: A rather plain roughcast three-storey building, described at its opening as having "a wide staircase, lighted by beautifully leaded windows". Latterly the base of the *Royal Naval Association*.

See *Eakin; Spectator 30 July 1932.*

PROSPECT ROAD

Road from Hamilton Road to High Street. It existed as a lane in 1833, but was not developed till the later 19th century.

See *Seyers p.8.*

No.1: Parkview: *c.*1910: Two-storey house with half-timbered gable overlooking Ward Park; shouldered gable chimney; Arts and Crafts windows

including tripartite window at first floor.
See *Lawrence 11234, 11236*.

Nos.3-5: Gronville: 1902, by F W Lang for Capt T Ferguson: Pair of three-storey brick semi-detached houses with crested clay ridge, corbelled chimneys, red sandstone lintels and arches, and tiled dadoes to porches; windows altered.
See *App 50*.

Nos.9-21: Eureka Terrace: *c*.1895: Terrace of two-storey paired stucco houses with two-storey canted bays and slightly taller end house; no original windows or doors left.

Nos.23-25: *c*.1910: Pair of detached stucco double-fronted houses with ground floor canted bays, and sawtooth bargeboards.

Nos.35-53: Verington Terrace: *c*.1905: Terrace of three-storey stucco houses with two-storey canted bays, nos.47-51 with segmental-headed windows, and no.53 larger.

No.2: *c*.1895: Two-storey double-fronted unpainted stucco cottage with finials on wallhead gablets. Four-panel door with fanlight and sidelights; first floor windows segmental-headed. No original windows.

Nos.4-24: *c*.1910: Terrace of two- and three-storey stucco houses with paired doorcases. The gap in the terrace at no.26 filled 1997.

Nos.28-36: *c*.1895: Miscellaneous terrace of two- and two-and-a-half storey stucco houses.

Q

QUAY PLACE: See *Crosby Street*

QUAY STREET

Deriving its name from the former coal quay and the old piers which jutted into Bangor Bay from each end, this short one-sided street was originally built up on the seaward side as well with a row of buildings known as *The Parade*. This was built on part of the present road, which was widened in 1914, and faced up High Street.
See *BHS III pp.37-42; Eakin; Green pp.10, 28-9; Hogg 38, 39; Lawrence 2859, 4721, 4727, C6019, I1991, I2231, I2685; Seyers pp.7-8; Welch 6, 10*.

THE PARADE: This area, which is now part of the Marina development (see *Piers*) has seen considerable change over the years. The NEW MILL for cotton spinning was built here in 1806 by George Hannay (see Introduction). Powered by steam, it was five stories high and employed 115 men and 98 women in the 1830s, along with numerous people in cottages sewing and weaving. By the 1850s it was owned by Wallace & Co, when on 2 April

QUAY STREET

1856 there was a fire in the machine room, followed by another in November, and it was forced to close. Nearby was the STEAMBOAT HOTEL, run by Charles Neill till he sold it to Annie O'Hara in 1886; this was a three- or four-storey building four bays wide, with vermiculated quoins at the ground floor framing a long shopfront, and entablatures and semicircular hood mouldings with bosses above the small first floor windows. There was another pub, and William Smiley's blacksmith's shop beside the pier end. The mill ruins were cleared about 1892 along with neighbouring buildings of The Parade to form *The Esplanade*. This was enclosed with iron railings, and contained a bandstand (later moved to the Marine Esplanade, *qv*) and the Coates Memorial which still stands here. The McKee Clock was erected in 1915, and further alterations in the 1950s formed the *Sunken Gardens*, which in turn was absorbed in the Marina development about 1990. The Quay Street elements of the marina are dealt with here.

See *BHS III p.38; Crosbie pp.8, 9; Eakin; Lawrence I2231; McCutcheon pl.134.6; Milligan pp.14-16; NDH 9 Nov 1889; OS Mems; Spectator 1 Aug 1931, 15 Dec 1956, 15 Sep 1961; WAG 346, 3103; Wilson pp.7, 8, 21, 26, 28.*

171 **Coates Memorial:** 1893: This is a cast iron fountain with frilly Moorish arches ornamented with herons in medallions, sphinxes above consoles and singing birds round the base of the dome; and carrying the inscription "Erected by The Members of the Bangor Corinthian Sailing Club in memory of their sincere and true friend Mrs Arthur Hill Coates 1893". Mrs Coates was the first wife of the Royal Ulster Yacht Club's honorary secretary in the 1870s, and they lived at Seacliff.

See *Lawrence 4721, I1991; Pike p.72; Wilson p.7.*

171 **McKee Clock:** 1915, by Mr Bell: The McKee Clock was erected as an eyecatcher at the foot of High Street, following the removal of the bandstand to the Marine Gardens (qv) and James McKee's offer to the Council of £200 to erect a four-faced illuminated clock "in the vicinity of the Esplanade for the convenience of the public". McKee was the Borough Rates Collector, and with the rapid growth of Bangor in the latter decades of the century he had become a wealthy man. The clock tower was designed by the Borough Surveyor, built by John McNeilly of Victoria Street using stone from Ballycullen near Newtownards dressed by Thomas Blaney of Belfast. It stands on a quadrilobar column with pyramidal roof over the square clockfaces. Sharman D Neill made the works, and the faces were made of "skeleton iron, glazed with opal glass", intended to be illuminated at night by jets of gas. It was unveiled by Miss Connor in a ceremony on 8 July 1915, after which the company adjourned to the Grand Hotel for afternoon tea. Recently stone-cleaned, only two of the faces currently have hands.

See *Crosbie pp.8, 9; Lawrence I1991 etc; Milligan p.14; Spectator 20 Mar and 24 Apr 1914, 9 and 15 Jul 1915; WAG 346, 3103; Welch 6, 9.*

No.2: *c.*1860: Three-storey building with rounded corner linking High Street and Quay Street, until recently with curious Edwardian picture-frame shop

windows, now spoilt by box shutters. Stucco lettering over corner entrance.
See *Crosbie p.8; Eakin; Lawrence 4721, 4727, C6019; WAG 3103*.

No.6: Walker's Irish Linen: *c*.1900: Single bay four-storey gabled stucco building with shallow canted oriel window to first and second floors and small-paned upper sashes. This building housed the *Peveril House Hotel* in 1910, and the *Rockview Hotel* in 1920.
See *Crosbie p.8; Lawrence 4721, 4727, C6019; WAG 3103*.

No.10: The Palladium: *c*.1895, originally *Quay House*: Red brick building three storeys high capped by a pair of ornate Dutch gables with bull's eye windows. Built originally as E & W Pim's grocery emporium (they specialised in Tea, from Extra Quality at 3/8 down to Good Strong Sound Tea at 2/-), soon after with tea rooms in the upper floors, the building has now become an amusement arcade, picking up rather untidy first floor oriels on the way.
See *Crosbie p.8; Eakin; Lawrence 2367, 4721, 4727, C6019, I1991; Lyttle; Spectator 27 Nov 1986; WAG 3103; Wilson p.45*.

Nos.12-14: McDonald's: *c*.1880: Three-storey two bay stucco house and shop, painted bright yellow and red. In Edwardian times, this was the *Victoria Restaurant*.
See *Crosbie p.8; Lawrence 4721, 4727, C6019, I1991; WAG 3103; Welch 6; Wilson pp.7, 45*.

No.16: Petty Sessions: 1866, as the *Belfast Bank*: Two-storey five bay stucco building in Italianate Classical style, very sophisticated for the Bangor of its day. Round-headed windows to ground floor, rectangular at first floor with balustraded bases and bracketed pedimented tops; Tuscan doorcase with paired pilasters, quoinstones, bracketed cornice, hipped roof. Disabled access recently added at the front with no regard for the symmetry of the design. A site near here was occupied from 1741 by the PRESBYTERIAN MEETING HOUSE, till the congregation built the church in Main Street. When the bank moved to the former market house, the court left the Dufferin Hall and took up residence here.
See *Crosbie p.8; Hogg 43; Lawrence 2367, 4721, 4727, C6019, I1991, I2231; Lyttle p.37; Spectator 24 Sep 1955; WAG 3103; Wilson pp.7, 45*.

Nos.18-20: Marine Court Hotel: 1993-94, by Derrick White for Eamonn Diamond: Five-storey five bay rendered building with attic in mansard floor; ground and first floors linked in darker granite podium. Balconies to third and fourth floor windows. Built by Savage Brothers.

This "beautiful and bracing position on the Esplanade" was formerly occupied by one of Bangor's most idiosyncratic Victorian buildings, the GRAND HOTEL, built about 1895 with crowstepped gables, little dormers, and conical-roofed corner turrets flanking ornate cast iron balconies, the very stuff of seaside towns. An interesting photograph in Bangor Heritage Centre shows the Hotel being erected, the right-hand turret finished and open before the other half of the site was begun. It was developed by Mrs Annie O'Hara, who had acquired the Steamboat Hotel on the Parade opposite shortly before it was demolished

QUAY STREET

along with the New Mill buildings. She then acquired two-storey stucco buildings which had formerly been sea-captain's HOUSES, on this site. She developed her new premises rapidly, advertising it as "The Largest and Best Appointed Hotel in the Town", complete with sumptuous rooms for smoking, sitting, reading, writing and playing billiards. When Mrs O'Hara went bankrupt (the coal dust from Neill's pier settling on the linen tablecloths was blamed) and retired from the business in 1910, the contents, including three "upright grand" pianos, were auctioned, and the den of iniquity itself was acquired by a syndicate of "local and Belfast gentlemen who are interested in the cause of temperance". Their aim was to buy up every licensed house that came on the market in Bangor in order to make it into a second Bessbrook. (They would be sadly disappointed to see the town today). When the Picture House opened in Main Street, the Grand Hotel offered a "High Class Cinematograph Entertainment" in its ballroom. For some years it was run as a home for underprivileged boys, till it was acquired in 1927 by another strong-minded lady, Mrs Louisa Barry.

As "Barry's", the building took on a new lease of life: it was famous for its dodgems and other fairground amusements, not all of which were appreciated by the neighbours in Albert Street. In June 1949, Mrs Minnie Delino, Mrs Barry's daughter, promised to stop the "Cage of Death", and was trying to reduce the noise of the "Demon Whirl". Among the most memorable of the amusements was a collection of extraordinary Edwardian penny-in-the slot tableaux including a Haunted House, and Executions of various nations. In 1982 Barry's was sold. Following a controversial spot-listing by the Department of the Environment, which was rescinded at the insistence of a slight majority of North Down Borough Council, the building was demolished early in 1984.

See *BHS III p.29; Crosbie p.8; Eakin; Hogg 42, 44, 46-48; Lawrence 2367, 4721, 4727, C2356, C6019, I1991, I2231, I2685; NDH Feb 1889, 24 May 1889; Spectator 25 Nov 1910, 5 Apr 1912; 11 June 1949, 15 Sep 1961, 5 & 19 Jan 1984; UA Aug 1994 pp.16-17, Jan 1995 pp.17-19; Ulster Tatler Nov 1984 pp.126-127; WAG 3103; Welch 10; Wilson pp.7, 27, 28, 45.*

No.24: The Windsor Bar: *c.*1900, probably by J G Lindsay: Three bay five-storey stucco building with central bay canted and topped by a gable which has recently been joined by outer companions. The site was previously occupied by two-storey HOUSES that became the *Harbour Bar*, a plain stucco terrace building. The present building originally opened as the *Burlington Restaurant*, complete with stucco enrichments since removed, and offering Dinners and Teas from 3d. With the arrival of the cinema, a *Picture Palace* was opened in the rear of the premises, and in 1915 the Irish Electric Palaces Ltd moved their *Picture House* operation from Main Street down to this building, opening that March. For the purpose, the building was "painted a pleasing French grey with suitable electrical illuminations" including "two powerful arc lamps of one thousand candle power each". Inside the entrance hall, however, was "the restfulness of a sylvan glade", even the pay office

Quay Street: a fascinating photograph showing Edwardian Bangor emerging from the smaller early 19th century buildings. Particularly interesting is the literally half-built Grand Hotel which had yet to acquire its left hand part. (NDHC).

Two structures from the former Esplanade (see Quay Street): on the left, the Coates Memorial of 1893, a cast iron drinking fountain; and on the right the McKee Clock of 1915 at the foot of High Street. (Peter O Marlow).

34 Quay Street: Bangor's oldest building is the 1637 Custom House. Originally it stood boldly on the rocky foreshore of the Bay but now it is skirted by cars rather than waves. (Peter O Marlow).

being rusticated. The new Palace could seat up to seven hundred patrons, and the screen was reckoned one of the largest in Ireland. On other nights, music hall artists such as Charles Coborn performed on stage, and in the adjoining Arcadian Gardens "first class concert parties" were given, until in 1940 the Picture House was gutted by fire. It never re-opened as a cinema, but Barry's amusement park took it over for its Ghost Train. It became a pub about 1960, expanding in 1986 to include a new function room that incorporates railings from Belfast's old Grand Central Hotel.

See *App 11, 996; BT 30 May 1986; Crosbie p.8; Hogg 42, 44, 46; IB 15 Oct 1899 p.156; Lawrence 2367, 4721, 4727, C2356, I1991; Spectator 28 Sep 1940, 17 May 1963, 24 April 1986; Ulster Tatler Nov 1984 pp.129-30; Welch 10; Wilson pp.7, 27, 45.*

Nos.26-28: The Royal Hotel: 1931-32,by L H Hodgins for Mr O'Hara: Five-storey rendered building, six bays wide with six-storey corner turret (now vying for attention with some rooftop oil tanks) and giant pilasters with Art Deco ornamentation between each bay. The inscription on the gable below a lion rampant, "Built by Iame Lyons in the year of Our Lord 1773" presumably comes from the plain three-storey ancestor of the present building (although Harry Allen has drawn attention to a hotel in Donaghadee that was owned by James Lyons in 1840), which appears to have been enlarged about 1895 by converting the small windows of the second floor into a pair of dormer gables. A little two-storey building to the south was also incorporated in the present building. The Royal was originally established about 1840 by a very active local figure called Henry McFall, who was also the harbourmaster, an insurance agent, poor law guardian and secretary to the Bangor Gas Company, as well as a keen swimmer. In 1908, the Royal was bought by William O'Hara, whose mother was about to close the adjoining Grand Hotel, and his family ran it until the mid 1990s.

See *App 2809; BHS III p.41; BT 15 Aug 1995; Crosbie p.9; Eakin; Hogg 43, 44, 46; Lawrence 2367, 4721, C2356, I1991; Minutes Dec 1870; Spectator 9 July 1932, 4 Jan 1958, 8 Sep 1961; WAG 346; Welch 10, 12; Wilson pp.7, 27, 30, 45.*

Nos.30-32: The Steamer Bar: *c*.1860 and later alterations: Solid four-storey building, formerly stucco. In 1864 it was the three-storey *Abercorn Hotel*; ten years later it was the picturesquely-named *Mermaid Hotel*. It reverted to the earlier name, but about 1910 became the *Marine Hotel*, and prosperity had added an extra storey, pedimented windows and octagonal pots: the fashion of more recent years subtracted most of these features, changed the name to the *Marine Bar* and added picture windows. With the present change of name have come some tram-carriage-like oriels at the second floor and a ground floor extension.

See *BHS III p.39; Crosbie p.9; Eakin; Lawrence 4727; Morton p.31; Spectator 8 July 1993; WAG 346; Welch 10, 12; Wilson pp.7, 30.*

No.34: The Custom House: 1637: This split-stone rubble building with limestone dressings to openings and quoins is the only tower house in the province now to occupy an urban site, and it is a most important landmark

QUAY STREET

along Bangor's seafront. As Mr & Mrs Hall reported in 1840, it is "still in good condition, and retains tokens of huge strength".

It can be accurately dated to 1637, when Charles Monck reported on the Customs in the northern parts of Ireland and wrote that "There is a fair custom house built but not finished by the Lord of Clanneboy, who hath received between two and three hundred pounds of the King towards it, and hath bestowed at least six hundred pounds already and two hundred more will hardly finish it". Monck describes the "large pile of stone made with flankers... very large storehouses, lodging chambers for officers, with chimneys, studies, and places to lay all sorts of commodities in", and declares that "if it were finished it were the best custom house in Ireland".

Although to a casual glance it appears to be two buildings, a stone tower at the corner and a Georgian house alongside, the House is one two-storey building some 40 feet by 20 feet in plan with a crowstep gable to the north, a four-storey battlemented corner tower at the NW, and a quarter-round staircase projection starting alongside it from first floor level. Three original lights have been identified in the brick wall at that level, but the main elevation has been considerably altered, with sash windows inserted possibly as early as 1837 when Lewis noted the "old building supposed to have been used as a custom-house, the tower of which has been converted into dwelling houses". Indeed it did not serve long as a custom house, since it was leased in 1672 by the owner of a corn mill, and by 1744 Harris noted it as "an oblong Pile of Building with a Tower at the North End, which is now in Ruins".

A nautical connection remained during the 19th century when David Harvey "the late popular and justly respected Harbour Master of the port" occupied it, though by 1881 the tower had become the summer studio of "Mr Robert Seggons, the celebrated Photographic artist". In 1923 Lady Clanmorris sold the tower to Bangor UDC, and in 1933, hot salt and fresh water baths opened at The Tower House, along with a suggestion that "the Old Tower be removed for road-widening purposes". Although "guaranteed to relieve rheumatism, sciatica and other disorders", the baths closed in 1954 because they were losing money, and latterly it was an antique shop. Following restoration by Larry Thompson & Partners for North Down Borough Council, it re-opened in 1983 as an excellent tourist information and exhibition area.

Thomas Morgan said that before about 1880, "there was not a foot of seawall between the Tower and Ballyholme", and only a cartway along by the Tower. An old lady brought up in the Tower House about the turn of the century recalls listening to the splash of the sea first thing in the morning, and how her mother could predict the weather for the day from the cries of the seagulls; watching the crowds streaming to and from the *Slieve Bernagh* twice a day; and how in the winter "the waves swept over the house and landed at the back of it". With the new breakwater these days are now past, but it is not so long since the Custom House was right on the sea.

See *Arch Surv pp.227-28, pl.64; BM Hartley 2138; Crosbie p.10; Eakin; Hall III*

QUAY STREET

p.20; Hogg 49; Lewis I p.183; Lyttle p.82; Milligan pp.38, 55; Old Custom House; MS by Anna Leonora Corkill; PRONI T.615/3; Spectator 10 Oct 1931; WAG 328; Welch 10, 13; Wilson pp.30, 49.

Harbour Master's House: *c.*1860: A small square-plan rubble-stone building with recently replaced sandstone dressings, hipped roof and corbelled brick chimney, on the sea side of the road and opposite the Custom House; but it is no longer by the sea, from which it was severed by a car park about 1990. Apparently a single-storey building from the road, it actually has a basement on the other side which was formerly a boat-house complete with slip.
See *Crosbie p.10; Eakin; Hogg 49; Lawrence; Seyers p.7; WAG 328; Welch 12-14; Wilson pp.22, 30, 49.*

QUEEN'S PARADE

Single-sided street along the original shore of Bangor Bay from the bottom of Main Street to join the Marine Esplanade. A fair bit of the old sea wall and weathered sandstone coping survives on the sea side. Known at the beginning of the 19th century as *Shore Street* and later as *Sandy Row*, this was possibly the *Front Street* listed in the 1842 Directory. The street was renamed after a visit by King Edward VII and Queen Alexandra (who "was quite lame and used an umbrella or stick to help her along"). The extension of the Parade from Gray's Hill towards Pickie is known as the *Kinnegar* from a coney or rabbit warren shown occupying that area on the Raven Map. A lively pair of naïve paintings now in North Down Heritage Centre depict Shore Street in 1842 and the Kinnegar in 1860 as recollected by Thomas Hanna of Gray's Hill. Until the mid 1980s, Queen's Parade was the only area of Bangor's seafront that had changed considerably since before the first war, and the almost completely residential character of 1900 had gone along with many of its buildings. Lack of investment during the development of the marina has been followed by compulsory acquisition of properties on the main Parade, which are currently demolished or boarded up awaiting comprehensive redevelopment.

177, 178

See *Crosbie pp.26, 27; Eakin; Hanna; Hogg 36, 37; Lawrence 2363, 2864, 3878, 3879, 4739, 4740, 5479, 1557, 12683; Milligan p.14; Seyers pp.12, 13; Spectator 22 Feb 1996; WAG 3104; Welch 3, 5, 7, 26; Wilson pp.8, 10, 26, 36, 40.*

No.1: See *2 Main Street*.

No.2: The Fountain: *c.*1900: Single bay three-storey building with Dutch gable, re-rendered without noticing the second floor window in the process. Replaced a single-storey house set forward from the present building line, which had its Dutch gable to the street.
See *Hanna; Lawrence 5479; Wilson pp.8, 26.*

No.3: Queen's Parade Methodist Church: 1891: Gable-fronted church of random squared basalt with nicely weathered sandstone trimmings flanked by pilasters terminating in lotus-like finials. Central doorway below paired pointed windows, each with a six-lobed window. It was built by William

QUEEN'S PARADE

James Campbell at the same time as the separate hall in Main Street (*qv*). It replaced a Methodist MEETING HOUSE and hall combined which was stucco-fronted and primitive Gothic in design, and had been built for a cost of £600 in 1820. That original church was built by the Wesleyan Methodists, who seem to have declined rapidly in numbers till in 1828 the 2nd Presbyterian congregation was worshipping in their vacant meeting-house. (This decline may have been because the Methodist leader, Rev Matthew Langtree, had become a total abstainer, not initially a popular move: when the Wesleyan conference in due course decreed that its ministers must abstain from alcohol, one participant said that he had enjoyed the conference "only fairly, you know it was very dry"). The New Connexion Methodists seem to have taken the church back in 1835.
See *Haire, passim; Hanna; Lawrence 4739, 5479, I557; OS Mems p.24; Presb Hist pp.113-15; Seyers p.12; Wilson pp.8, 26.*

No.4: *c.*1995: Three-storey rendered building with channelled ground floor and central glazed feature. Previously this was a three-storey four bay stucco HOUSE with carriageway entrance to which a curved oriel window had been added, which had latterly been a café, then an amusement arcade.
See *Hanna; Lawrence 5479, I557; Wilson pp.8, 26.*

Nos.5-9: Papa Capaldi, Grahame, Trumps, Vacant, Pizza King: *c.*1885, possibly by J V Brennan: Terrace of four three-storey stucco houses with triplet windows above first floor castellated bow oriel windows. Shopfronts have destroyed the wineglass stems that originally supported the oriels as pilasters at ground-floor level. Hanna shows a single-storey cottage with clock in a blind dormer as being on part of the site in 1842. In the 1950s, the *Queen's Cinema* was located down an entry in this terrace.
See *Hanna; Lawrence 5479, I557; Wilson pp.8, 26.*

No.10-12: Paul's Burgerbar, Four A Cabs, The Sweetie Jar: *c.*1880: Terrace of two-and-a-half storey stucco buildings, each with two frilly dormers, some original sashes, but marred by rather garish shop fronts. This may be *Corazon Houses* which were listed in the rates books for 1872, or *Nora Villas* listed about 1910.
See *Hanna; Lawrence I557; Wilson pp.8, 26.*

Nos.13-14: Marina Bay Leisure, John Dory's: *c.*1830: Three-storey four bay stucco building, originally a pair of rather dignified houses with channeled ground floor and Ards doorcases; even restoration of its chimney-stacks would help restore its grandeur.
See *Hanna; Lawrence I557; Wilson pp.8, 26.*

Nos.15-25: *c.*1935: Terrace of three-storey smooth-rendered buildings with shallow canted bays and attic dormer, with shops at ground floor. In 1953, the mayor opened Queen's Parade Shopping Arcade at no.25, a "brilliantly illuminated corridor, flanked by attractive lock-up shops". Now mostly vacant, awaiting redevelopment; no.25 has been demolished. Hanna shows a neat terrace of half a dozen two-storey houses on this site, all with Ards doorcases,

Queen's Parade in 1842, as painted by Thomas Hanna. There is still a pub at the left hand corner, and a Methodist church near it, but nearly all the structures have been rebuilt. (NDHC).

5-9 Queen's Parade, photographed in 1968: before the insertion of shopfronts the castellated oriel windows of this c.1885 terrace had elegant wineglass stems. (H A Patton).

The Kinnegar end of Queen's Parade photographed in July 1874: many of the houses on the Kinnegar and at the foot of Gray's Hill are single-storey cottages, and there is no sign of tourist activity! (NDHC).

View from the pier along Queen's Parade about 1875: in the background can be seen the much grander early development of Princetown on the low hills above Bangor Bay. (NDHC).

and some with entries.
See *Hanna; Lawrence 1557; Spectator 4 July 1953; Wilson pp.8, 10, 26, 40.*

Nos.27-30: Gap site: This was formerly the REGENT PALACE HOTEL of 1931, by L H Hodgins, a four-storey steel-framed building with very shallow oriel windows which cost £20,000. It later formed part of the *Queen's Court Hotel.*
See *BNL 27 June 1931; Eakin; Hanna; Hogg 37; Spectator 4 July 1931; Wilson pp.10, 40.*

No.31: Caesar's: *c.*1900: Three-storey stucco building with picture windows and plastic shopfront; second floor windows linked by lozenged string course.
See *Hanna; Wilson pp.10, 40.*

Nos.32-34: Vacant: *c.*1850: Two-storey four bay building with upper floor now pebbledashed; Gibbsian surrounds to first floor windows.
See *Crosby p.26; Hanna; Wilson pp.10, 40.*

Nos.35-38: Art & Gift Shop, Vacant, Wong's Restaurant, B J Eastwood: *c.*1880: Terrace of three-storey stucco buildings with paired round- and basket-headed windows in shallow oriels.
See *Crosby p.26; Hanna; Lawrence 4740; Wilson pp.10, 40.*

No.39: YMCA: *c.*1920: Three-storey rendered building with long oriel window jutting out at second floor, decorated with sinuous leaded glass. This was originally built as Enrico *Caproni's* splendiferous ice-cream emporium, "Unrivalled for Variety & Delicacy of Flavour".
See *Crosbie p.26; Hanna; Wilson p.10.*

Nos.40-41: McBurney's Bar: *c.*1910: Irregular and rather eclectic design three storeys high with octagonal corner turret decorated with sun-rise motifs; formerly the *Strand Hotel*, and latterly the *Warwick Bars*. Nos.1-9 Southwell Road are contemporary. Previously the houses on this site were stucco and two-storey. James Crosbie had *hot salt water baths* here, pumping the water from the sea with a small force pump.
See *Crosbie p.26; Hanna; Lawrence 4740, 12775; Seyers p.2; Welch 3, 5, 7; Wilson p.10.*

No.42: Kinnegar Guest House: *c.*1910: Double-fronted three-storey stucco house at end of terrace with two-storey canted bays. This is the start of the *Kinnegar* terrace, houses developed at the turn of the century with long front gardens, in place of an irregular row of mostly single-storey cottages that overlooked "a broad grassy bank, as well as a roadway, over which you could look down at the waves dashing against its base at high tide." This was actually the last of the Kinnegar houses to be redeveloped and the site was still occupied about 1900 by a single-storey COTTAGE with label mouldings over its windows.
See *Hanna; Lawrence 4739, 12775; Spectator 22 June 1906; Wilson pp.10, 36, 40.*

Nos.43-46: 1902, by Samuel Stevenson for George Matthews: Two pairs of three-storey stucco houses with three-storey canted bays, nos.45-46 somewhat altered. Built by Robert Neill and finished after his death by Lester Irving.

QUEEN'S PARADE

The HOUSE formerly on the site, which Hanna shows as a substantial two-storey four bay one with off-centre Ards doorcase, quoins, labels over windows and a panelled parapet, was occupied by Mr Kennedy, surveyor to the agent of the Bangor estate, and afterwards secretary to Lord Dufferin
See *App 69; Hanna; Lawrence I2775; Seyers pp.13, 35; Spectator 22 June 1906; Wilson pp.10, 40.*

Nos.47-56: Beaumont Terrace: *c*.1890: Terrace of three-storey stucco houses with full height canted bays, ornate keystones over ground floor windows, and handsome terracotta balustrading along the esplanade. Most retain their sash windows and panel doors. About here was the single-storey COTTAGE of Thomas Whannell, whose two daughters married builders working on Bangor Castle (qv) who went on to greater things in Australia; next door to them lived a tallow chandler.
See *Crosbie p.27; Hanna; Hogg 36; Lawrence 3879, 11217, 11633, C6010, I2775; Spectator 22 June 1906; WAG 3104; Welch 7, 26, 27; Wilson p.40.*

Nos.57-58: 1883: Pair of three-storey gable-fronted stucco houses, partly pebbledashed; row of four minute sashes in centre at second floor.
See *Crosbie p.27; Hanna; Hogg 36; Lawrence 3879, 11217, 11633; Welch 26; Wilson p.40.*

Nos.59-62: Emmaville: *c*.1885: Terrace of three-storey stucco houses with two-storey canted bays, nos.61 and 62 with iron balustrading to bays. The single-storey cottages previously on the site, on the edge of the mid-19th century town, were set further back and had long gardens down to the sea.
See *Crosbie p.27; Hanna; Lawrence 11217, 11633; Seyers p.13; Spectator 21 Jun 1906; WAG 3104; Welch 5, 26; Wilson p.40.*

Mount Pleasant: For this and the remaining terraces along the Marine Esplanade see *Princetown Terrace, Mount Pleasant, Mount Royal, Princetown Road, Pickie Terrace* and *Lorelei.*

QUINTIN AVENUE

Short road from Gransha Road to Demesne Avenue, laid out about 1925 but redeveloped *c*.1970 as Sunningdale Park.

R

RAGLAN ROAD

An attractive winding road on a hillside rising from Princetown Road to Downshire Road, with mature gardens. The southern portion of the road was in existence by 1833, when it was part of a lane leading to a farm, but the road as a whole only developed around 1885.

No.3: *c*.1890: Three-storey unpainted stucco house with two full-height canted bays, similar to the Mornington Park development at 103 Princetown Road.

Gate pillars with cement scrolls match those at Mornington Park.

Nos.5-7: Tyneholme: 1903-04, by J J Phillips & Son for C E Dyer: Two-storey stucco semi-villas with canted and rectangular bays and stucco plaques carrying the name and date.
See *App 158.*

No.2: The Rectory: *c.*1890, formerly *Raglan Lodge*: Two-and-a-half storey double-fronted villa with two-storey canted bays below curly Dutch gables; fluted pilasters to door.

No.4: *c.*1900: Two-storey stucco double-fronted house with two-storey canted bays; crested clay ridge, fretted apex board to side gables; stained glass fanlight and sidelights to front door.

No.6: *c.*1925: Two-storey red brick house with hipped bellcast roof and cast iron rainwater hoppers; strange cubic lumps on chimney-stacks.

Nos.8 and 10: *c.*1900: Pair of two-storey double-fronted stucco houses with two-storey canted bays. Moulded surrounds to segmental-headed plain sash windows.

Nos.14-16: Marine Villa and **Sunnyside Villa:** *c.*1880: Two-and-a-half storey stucco semi-villas with ornate bargeboards and apex boards, and dentilled cornice; windows in strap surrounds. No.14 has the name *Marine Villa* on its gate pillars.

RAILWAY VIEW STREET

A steep hillside street rising from the station, the south side of which was developed during the late 19th century with stepped terraces of two-storey houses. Nos.2-12 were called *Hillman Street* in 1920.
See *Lawrence 9544; NDH 28 Aug 1888.*

Nos.1a-7a: *c.*1875: An entry behind no.1 leads past some ramshackle railings of old railway sleepers to a group of tiny houses at the bottom of a lane. No.3a is of lined stucco with margin-paned sashes and four-panel door.

Nos.1-45: *c.*1880: Terrace of two-storey painted stucco houses, stepped up the hill in pairs. No.1 still has original sash windows and four-panel door with an escutcheon for painting in the street number. Nos.1-9 (and **nos.55-61** further up) have a high wallhead with string course above first floor windows. **Nos.47-75** (*Emerald Terrace*) are broadly similar in date and type.

Nos.77-81: Ralph's Hill Terrace: 1888, by Robert Robinson: Three-storey stepped stucco terrace with two-storey canted bays, with name and date on stone plaque. It is not clear who Ralph was, but his hill was nearby on Manse Road and named on the Raven map of 1626.
See *BHS I p.61; Lawrence 9544.*

Nos.2-20: *c.*1900: L-shaped terrace of two-storey rendered houses. Front doors have sidelights on one side only, and first floor windows are alternately very narrow.

RANFURLY AVENUE
This street was laid out from Farnham Road to Maxwell Road about 1905.

No.17: *c.*1925: Two-storey red brick house with stained glass upper lights to tripartite windows; Diocletian window at first floor; rosemary-tiled verandah; bevelled light to door.

No.21: *c.*1930: Two-storey roughcast house with quoins; narrow sashes in rows with leaded lights.

No.43: Laurel Lodge: 1900-01, built by H Seaver for Charles Lepper: A substantial two-storey red brick house with pinkish sandstone dressings and pediments; a rather severe classical design relieved by brackets to cornice and pediment, and by tall brick chimneys with stone cappings. Built as a private house, it was *Garth House* school for about forty years, and entered from Maxwell Road, but when it ceased to function as a school about 1965 its grounds were developed, and the house has been converted to flats.
See *App 19.*

No.8: Netherleigh: *c.*1905: Two-storey double-fronted pebbledashed house with rosemary-tiled roof; a simple late-Edwardian house.

No.38: *c.*1900: Irregular two-storey stucco house with corbelled stucco chimney, bargeboards ornamented with the vestiges of finials, and a mixture of window types dominated by paired round-headed windows under double hood-mouldings at first floor.

No.40: *c.*1900: Two-storey pebbledashed house, heavily becreepered. Slate roof with slightly corbelled red brick chimneys with tall red pots, eaves with projecting rafters. Casement windows with small-pane upper lights. **No.42** has some similar details.

RATHGAEL ROAD
From the Belfast Road to the Newtownards Road, the Belfast end of the road was a realignment carried out in the 1850s to accommodate the lake at Clandeboye, but otherwise its track was present in 1830. A sort of outer ring that was in countryside not so long ago; much of it skirts the well-wooded Clandeboye estate, but increasingly the rest is becoming modern suburbia: or as the translation of the old Irish name *Rathgill* would have it, "fort of the foreigner".
See *Seyers p.45.*

Nos.15-37: Rathgael Villas: *c.*1950: Pairs of two-storey semis, several with slightly jettied mutual half-timbered gables.

RATHGAEL HOUSE: Before its demolition in 1960 (which used 115lbs of gelignite) to make way for a training school, this was a pleasant two-storey symmetrically-designed house five bays long, of harled split-stone rubble; there was a tripartite window at first floor above the large round-arched doorcase which contained round-headed sidelights, the fanlight being glazed

in spider-web pattern. Some of the windows and doors were apparently re-used at the Dunadry Inn, Templepatrick. The house was 18th century in date with some 19th century additions, and was the home of the Rose Cleland family, who founded the first Sunday School in Ireland there, for "children of either sex and of any religious denomination". In the 19th century, Richard Rose Cleland kept a pack of hounds, and when riding used to tie his long beard up to keep it from blinding him. There was a GATE LODGE, also demolished in 1960, which had been built about 1880, Mrs Rose-Cleland tracing its shape out in the snow one winter, "the builder pegging it out as he followed her". The name of the house has passed to a rather undistinguished building nearby which also concerns itself with education.
See *Arch Surv pp.381-82; Knox p.548; Lewis I p.183; Seyers p.46; Spectator 26 Aug, 2 and 16 Sept 1960.*

Ava Cottage: pre 1833: Former agent's house heavily obscured by trees. Was named *Ballylangley* in the 1930s but has reverted to its former name.

RATHMORE ROAD
From Crawfordsburn Road to Old Belfast Road.

West Presbyterian Church: 1963, by W McK Davidson: Strangely roofed church on rustic brick base with louvred clerestorey - a tall shingled roof with sloping ridge falling from front to back; north end has floor to ceiling coloured glass window, with roof overhanging beyond it. It makes a striking landmark: at its opening, the architect said, "At least if they don't like the church, they can't possibly ignore it."
See *Spectator 11 October 1963.*

Riverside Terrace: See 1-19 Springfield Avenue.

ROSLYN AVENUE
From Demesne Avenue to Jordan Avenue. Listed originally as *Rosslyn Avenue* in 1925.

Royal Terrace: See 188-194 Seacliff Road.

Royal Ulster Yacht Club: See 101 Clifton Road.

Royston Terrace: See 62-80 Southwell Road.

RUBY STREET
Short single-sided street from Hamilton Road to Castle Street, looking towards St Comgall's Church, probably developed about 1885.

Ruby Lodge: *c.*1885: A picturesque two-storey stucco house with projecting rafters, cockscomb ridge and skews and ball-finials to gables. This was formerly the schoolmaster's house.

Nos.3-25 and Castle Arms, Castle Street: *c.*1895: Terrace of stepped two-storey stucco houses with prominent string course below first floor windows, facing St Comgall's Church.
See *App 28; Seyers p.33.*

RUGBY AVENUE
Street developed during the 1930s on ground owned by Bangor Rugby Club, bending right from Brunswick Road and over a hill, down and on to link to Grange Road. Typically a mix of hipped-roof bungalows and two-storey semis, mostly pebbledashed with cement trims, but almost all now with plastic windows.

Nos.9-11: *c.*1935: Two-storey hipped pebbledashed semis with square red brick bays rising to bargeboarded gable over first floor Diocletian window; door recessed in brick surround; stained glass top lights.

No.23: *c.*1935: Roughcast bungalow with red tile roof and two dormers; square bays with stained glass top lights.

Ruthville Terrace: See 8-14 Albert Street.

S

St Ruan: See 42-48 Seacliff Road.

Sally Beattie's Hole: See the Long Hole, Seacliff Road.

SANDHURST DRIVE
Off Sandhurst Park. Laid out around 1930.

SANDHURST PARK
Off Groomsport Road. Laid out about 1925, and complete by 1935.

SANDYMOUNT
Cul-de-sac off Quintin Avenue parallel to Gransha Road, laid out around 1930, but redeveloped *c.*1970 as *Sunningdale Park.*

SANDRINGHAM DRIVE
Road from Ballyholme Esplanade to Groomsport Road, developed during the 1930s on former football grounds. Typically two-storey semis with gabled two-storey bays, often in red brick with pebbledashed first floors.

Nos.2-4: *c.*1935: Pair of Modernist flat-roofed rendered two-storey semis with central curve-sided bays and corner windows at first floor.

SANDRINGHAM GARDENS
Cul-de-sac off Sandringham Drive developed during the 1930s, with houses arranged radially at the end.

SANDY ROW: Former name for *Queen's Parade*.

SEACLIFF ROAD
A very important stretch of late 19th century villas running round the coast at *Luke's Point* between Bangor Bay and Ballyholme Bay, with most of the houses on the inland side of the road and commanding superb views of the bay. Although the line of the road was present as a track in 1833, at which time the peninsula was dedicated to George rather than Luke, it only became a proper road to Ballyholme about 1860, after which it was rapidly developed. On the seaward side there is only occasional development, and most of the houses have unrestricted views across Belfast Lough. At first the seaward side of the road is lined with rendered stone walls, but after Seacliff there is an open iron railing. After Luke's Point, the rocks are covered by the debris of concrete sandbags and brick and reinforced concrete from the town's airraid shelters which were dumped there after the war, now covered with limpets and wrack.
See *Lawrence 2865, 2364, 3877, 4738, 6200, 11228, 11229, C5054; Welch 20, 21, 58.*

The Long Hole: An inlet between rocks to the north of Bangor Bay, and almost certainly to some extent a natural, although possibly very shallow, harbour. Apparently it was further excavated by Col Ward, who spent £20,000 to create the 19th century equivalent of a marina. It was sometimes known as *Sally Beattie's Hole* after the occupant of a thatched cottage that used to overlook it, and sometimes as *The Big Hole*. The author's grandfather, Marcus Patton, recalled as a boy crawling on his hands and knees past the old Custom House on one fiercely stormy day (probably in the 1894 gale), and watching a sailing ship tacking to and fro off the shore till it was eventually driven onto the rocks at the Long Hole. He could see the crew praying on deck, when a large wave lifted the ship into the Long Hole, and miraculously they were able to "walk" ashore.
See *Crosbie p.11; Eakin; Hogg 59; Lawrence 2364, 3877; Spectator 7 Sep 1906, 5 Sep 1931, 23 Feb 1957; WAG 2172; Welch 20, 21, 58; Wilson pp.29, 31.*

Nos.1-9: Seacliff: *c.*1780: Two-storey L-shaped block of stucco houses built out on the rocks, those on Seacliff Road ending as single-storey houses; simple details with quoins and broad-framed sash windows, but incised Greek-key pattern in doorcases suggests some sophistication at one time. The main house was also at one time called *Sea Cliff*, and at another as *Saltpans*, while the smaller houses were known in the 1930s as *Seacliff Cottages*.
See *Lawrence C5054; Seyers p.11; Welch 23, 58.*

Former Bathing House: *c.*1880: Alongside the RUYC Battery (built in

1-9 Seacliff Road: this L-shaped building on the rocky foreshore probably dates from about 1780 and was the earliest development in this part of Bangor.
(Peter O Marlow).

102-112 Seacliff Road: Clifton Terrace was probably built about 1857 to cater for the infant market in seaside holidays, which took off shortly after with the arrival of the railway. (Peter O Marlow).

1961), a low whitewashed rubble-stone building with barrel-roof, for gentlemen to dress in whilst sea bathing; there was formerly a similar structure at the Pickie Rock (see *Marine Gardens*).

Luke's Point Sewage Pumping Station: *c.*1935: An attractive one-and-a-half storey blackstone rubble building with hipped roof with dormers and tall chimneys, belying its utilitarian function.

Ballyholme Yacht Club: 1956 and 1963, by H A Patton: Irregular group of one and two storey buildings agglomerated over the years from a simple timber hut beside the battery on the rocks in 1937 to the present changing rooms, clubhouse and battery, and dominated by a two-storey central block with pitched felt roof. The original "dainty" CLUBHOUSE (1909, by Kennedy & Hill) was on the other side of the Seacliff Road and was roughcast with casement windows "fitted with rustic shutters, similar to those seen on the old English type of cottage"; it was comparatively small, and its main feature was a full-size billiard table that must have left very little room for yachting activities. The first stage of the present building was built on partly reclaimed land, using plate glass and "special electrical fittings... designed to be in keeping with the nautical atmosphere"; while the second phase, costing £7,600, included a "spacious observation lounge" on the first floor. A later extension was carried out in 1972, while the training building nearby (with a wrought iron yacht screening a gable porthole) was finished in 1997.
See *Spectator 16 Jul and 20 Aug 1909, 2 Jun 1956, 16 Sep 1960, 29 May 1964.*

Nos.2-6: Bayview: *c.*1890: Terrace of three-storey stucco houses with shops later inserted at ground floor; pilasters at ends of terrace, small chamfered depressed arch windows at second floor. Built by one Captain McCullough.
See *Eakin; Hogg 49; Seyers p.12; Welch; Wilson pp.22, 87.*

No.8: 1998: Block of three-storey "waterfront luxury apartments", in red brick over a channelled rendered plinth, with blue plastic windows. Until about 1985 this was the site of a single-storey stucco COTTAGE painted cream and green, with central porch, label mouldings to windows, quoins, triangular oriel window on one gable, and conservatory built on. It was built about 1890 and bought by the solicitor George McCracken as a wedding present for his wife, L A M Priestley, who was a relative of the novelist J B Priestley and used the cottage as a summer residence. She was a journalist who contributed to *The Strand Magazine* and other periodicals, and was the authoress of the racily titled *The Love Stories of Some Eminent Women*. She was also one of the first female cyclists in Ireland, causing quite "an embarrassing sensation" everywhere she went, a talented tennis player, and, from 1908 onwards, an advocate of votes for women: Sylvia Pankhurst and other suffragettes stayed with her on visits to Ireland.
See *Hogg 49; Spectator 19 Aug 1983; Ulster Tatler Jan 1984; Wilson pp.22, 87.*

Nos.10-12: Bangor Bay Inn: *c.*1895, formerly *Redcliffe*, for Dr R L Moore: Three-storey irregular red brick terrace on oblique-angled corner site,

ornamented with various types of bays, oriel, and cast iron verandah supported on delicate columns and acroterion spandrel pieces; bow window at No.12 terminating in conical roof. Previously there was a small once-thatched COTTAGE on this site, known after its occupant as *Sally Beattie's Cottage*.
See *Hogg 49; Lawrence 4738; Wilson p.22*.

Nos.14-18: Hibernia Terrace: *c*.1890: Terrace of three-storey stucco houses with two-storey bow windows, and corbelled aedicules over doors.
See *Lawrence 4738; Wilson p.31*.

Nos.20-36: Kerrsland Terrace: *c*.1890: Terrace of two-and-a-half storey stucco houses with two-storey canted bays, dentilled cornice and fretted apex panels to gables. Double-width bay at no.36. **Nos.38-40**, *Seaview Terrace*, are similar.
See *Lawrence 2364, 4738; Wilson p.31*.

Nos.42-48: St Ruan: *c*.1890: Terrace of two-and-a-half storey stucco houses with finialed gables above two-storey canted bays. Windows originally double-hung sashes, some with vertically divided sashes.
See *Crosbie p.11; Lawrence 4738; WAG 2172; Welch 21; Wilson p.31*.

No.50: Pakenham Villa: *c*.1885: Double-fronted two-storey stucco house with central glazed porch and canted bays at ground floor, quoins at sides; on a triangular site. When offered for let in 1892 it was described as a "beautifully furnished house... in one of the nicest portions of Bangor".
See *BNL 18 Oct 1892; Crosbie p.11; Eakin; Lawrence 6200, 13595; WAG 2172*.

No.52: Needwood House: *c*.1890: Drastically rebuilt *c*.1990, now much larger and with a clumsy three-storey portion. Originally a two-and-a-half storey double-fronted stucco house with paired windows above canted ground floor bays and central feature above doorcase.

Nos.54-58: *c*.1905: Irregular terrace of three-storey stucco buildings, no.54 having stained glass upper lights to casement windows.
See *App 66; Crosbie p.11; Lawrence 6200; WAG 2172*.

Nos.66-72: Knightsbridge: *c*.1905: Terrace of three-storey stucco houses with two-storey canted bays. Rubble-stone boundary wall.
See *Crosbie p.11; Eakin; Lawrence 4738, 6200, 13595; WAG 2172; Welch 21*.

Nos.74-80: Ailsa Terrace: *c*.1890: Terrace of three-storey stucco houses similar to nos.66-72, but with hipped roof, and originally with balustraded parapet. The terrace was presumably given its name because, as noted on a bye-law application for an adjacent site around 1900, these were "Mr Craig's houses". Ernest L Woods, the borough surveyor, lived at no.80 around 1910.
See *Crosbie p.11; Eakin; Lawrence 2865, 4738, 6200; WAG 2172; Welch 21*.

Nos.82-84: Clifdene: *c*.1990: Former gap site filled to the brim with three-and-a-half-storey red brick apartments on rendered plinth. Where earlier buildings disported themselves on the rising rock face above the shore, this building has cut into the stone.

Nos.86-96: *c*.1885-95: Varied terrace of two-and-a-half and three-storey

stucco houses, all with steep gardens, and flights of steps rising from random stone walls at the road. No.86 has a corbelled corner oriel at first floor level, and nos.94-96 have screen walls between their doors, which originally supported ornamental iron balconies. Built in stages, nos.94-96 being later than the others.
See *Crosbie p.11; Hogg 59, 60; Lawrence 2865; WAG 2172; Welch 20, 21, 23, 24; Wilson p.31.*

No.98: Seacrest: *c*.1895: Originally *Seacliff Villa*. Double-fronted stucco villa, two storeys high with verandah joining ground floor bays and with dormers in roof; windows unfortunately altered.
See *Crosbie p.11; Hogg 59, 60; Lawrence 2865; WAG 2172; Welch 21, 23, 24; Wilson pp.29, 31.*

Nos.102-112: Clifton Terrace: *c*.1857: Important group of two-storey rubble-stone houses set above the road with lawns behind roughly-coursed stone boundary walls. Stucco canted bay windows rising below gables; margin-paned windows; brick dressings to openings; many details altered, but well worth the effort of restoration. Reported in the Belfast Newsletter in May 1857 as being erected "in the very best manner, embracing all modern improvements in ventilation, and every convenience that can be desired", they were built by a Mr Cowan (possibly Andrew Cowan who was to acquire Glenganagh a decade later).
See *BNL 12 May 1857; Lawrence 2865, C5054; Morton p.29; Seyers p.11; Welch 20, 21, 23, 24; Wilson pp.29, 31.*

Nos.114-120: Edenville: *c*.1870: Simple terrace of two-storey stucco houses with basements, an unusual feature in Bangor; dentilled cornice, moulded surrounds to ground floor openings, pilastered entablatures to doorcases at the head of elegant railed steps spanning the basement areas; basement channelled. It was presumably one of these that was advertised to let in 1873 as "an elegantly furnished marine residence... at the New Houses adjoining Clifton Terrace".
See *Lawrence C5054; Morton p.30; Wilson p.29.*

Nos.122-144: Seacliff Terrace: *c*.1895: Terrace of three-storey stucco houses with two-storey canted bays. No.130 is taller, with a basement. **Nos.146-148** are semi-detached but were similar to the terrace before being altered.

Nos.150-160: *c*.1955, for Bangor Provident Trust: Two-storey apartments in rustic brick with first floor balconies. This site was late in being developed as it faces Seacliff rather than looking directly at the sea.
See *Milligan pp.34-35.*

Nos.162-170: Mossvale: *c*.1895: Terrace of three-storey houses with three-storey bow windows; segmental-headed vertically divided sash windows; pilastered and dentilled doorcases.

Nos.172-174: Clifton: 1855, possibly for David Connor: Pair of two-storey houses similar in date and style to *Clifton Terrace* but altered.
See *Seyers p.11.*

SEACLIFF ROAD

Nos.176-182: Belgravia: *c*.1900: Terrace of two-and-a-half storey double-fronted houses with two-storey canted bays, and gables finished with curious concrete tufts like Prince of Wales feathers. At the end of the terrace, **no.184** is a detached stucco house of similar date.

No.186: 1965, by H A Patton: Two-storey lopfronted house with balcony. Behind it, steep steps wind up to a weird structure of 1996 with portholes in a stone gable, and a brick turret without windows, that appears to be a sheer folly.
See *Spectator 28 Nov 1996*.

Royal Ulster Yacht Club: See *101 Clifton Road*.

Nos.188-194: Royal Terrace: *c*.1890: Terrace of three-storey stucco houses with three-storey bow windows.

King's Fellowship: *c*.1980: Flat-roofed two-storey structure in render and brick. This was the site of NARDINI'S MARBLE HALL, which then became *The Music Box*, *The Flamingo Ballroom* and finally the *Astor Cinema* - none of them quite as grand as their names might have suggested. The Music Box, for instance, boasted an "orchestra" of five musicians, and could seat five hundred.
See *Spectator 24 Jun 1939, 7 Nov 1942, 9 Aug 1947, 10 May 1963, 11 Sep 1964*.

Kingsland Park: In 1900, the Seacliff Road ended here, and Luke's Point was open rocks and whin bushes. However an ambitious switchback railway, described as "the longest switchback ride in the UK", was opened here in 1889 by Lady Clanmorris. It was destroyed in the 1894 gale, but the ground was acquired by Bangor Council in 1915 and laid out as a park. Barry's Amusements later opened here in the summer months. The first clubhouse of the Ballyhome Yacht Club was here. The public **toilets** at the foot of Ward Avenue in Arts and Crafts style (*c*.1920) are worth mentioning. Beyond it is a pitch and putt course and tennis courts.
See *Crosbie p.15; Hogg 4; Milligan pp.1, 4-5; WAG 400; Wilson p.16*.

No.244: The Bay: 1984: Three-storey flats of gloomy brown brick with recessed balconies set between spine walls, and skeletal gables over each bay; continues up Seaforth Road. Miami-style palm trees at front.

This was formerly the site of Enrico CAPRONI'S 1925 (extended 1935) *Palais de Danse* and *Mirimar Café*, which were closed in 1977 and demolished in 1983. The hipped roof structure was not exciting architecturally, but the sprung floor and white marble staircase played host to the dance bands of Henry Hall, Harry Roy, Victor Sylvester and many others, including Jimmy McDowell, the Shipyard Tenor. Whether it was for Lovely Legs Night, Scotch Week or New Year's Eve, *Cap's* attracted dancers from all over the province in its heyday, despite being teetotal. No doubt what drew them was the carefully phrased notice reading "Through these doors, from time to time, pass some of the most beautiful women in the world".
See *Crosbie p.16; Eakin; Hogg 4, 5; Spectator 27 Apr 1925, 27 Apr 1935; Ulster Tatler Mar 1984; WAG 2703*.

No.252: Kingsland Nursing Home: *c*.1990: Two-storey rendered building with polygonal bays set out at each end, on site of the former Bangor SHIPYARD.

Nos.256-262: The Ballyholme: *c*.1890: Terrace of three-storey stucco houses with conical roofs to full-height bow windows; paired doorcases with corbelled entablatures. Now a nursing home, it was used for a while (1900-1905) by *Dr Connolly's Intermediate School*, forerunner of the present Grammar School (see *College Avenue*), hence its original name of *College Gardens*, and until recently it was the *Ballyholme Hotel*.
See *Lawrence 11224, 11229*.

No.276: Seaview House nursing home: *c*.1912, probably by Ernest L Woods: Two-storey red-brick house with vertically divided upper sashes and dentilled eaves; considerably altered.

No.278: The Cairn: *c*.1910: Two-storey house with one end bay set forward as a gable terminating a verandah, with rough weatherboarding in gables; tall roughcast chimneys; rather Art Nouveau garden wall.

Nos.280-282: Ruskin Villas: *c*.1890: Three-storey stucco semi-villas with doors in outer bays; small duple windows in gables.

Seacliff Terrace: See 122-144 Seacliff Road.

SEACOURT LANE
Steep narrow lane from Princetown Road down to the Marine Esplanade, enclosed by walls some twelve feet high in rubble stone with occasional brick-dressed openings, overhung with trees which belong to the neighbouring properties of Glenbank and Seacourt (see *Princetown Road*).

SEAFORTH ROAD
Steep road from Seacliff Road to Ward Avenue, laid out about 1890.
See *Eakin*.

Nos.1-3: Ardmore: *c*.1890: Two-and-a-half storey red brick and pebbledashed semi-villa with gables to outer bays. Used 1897-1900 as *Dr Connolly's Intermediate School* (see *College Avenue*).
See *Lawrence 11224; Milligan p.52*.

Seaview Terrace: See 38-40 Seacliff Road.

SECOND AVENUE
From Ballyholme Road, linking to Fourth Avenue at the bottom of Fifth Avenue as part of the *Bay Lands* development. Development commenced about 1924 and was completed in the mid thirties.

No.5: *c*.1930: Half-hipped house with red artificial slate roof and brick gable chimneys. Ground floor red brick with central door in arched recess, first floor roughcast with windows rising into dormers. Nos.7 and 9 are of similar design but altered.

SECOND AVENUE

No.2: *c.*1930: Bungalow with half-hipped roof and two octagonal hipped dormers; bulgy columns support overhanging eaves of porch.
See *BT 9 Apr 1998*.

No.4: *c.*1930: Tiny hipped-roof roughcast bungalow at the top of a steep irregular mound of a garden.

Nos.6-8: Orlington and **Abbotsford:** *c.*1924: Pair of roughcast semis with mutual brick chimney-stack. Splendid sycamore and healthy horse chestnut in the garden of no.8 do much to make this avenue attractive.

SEVENTH AVENUE: See *Hazeldene Drive*.

SHANDON DRIVE

Road from Clifton Road down hill to Seacliff Road, with a view towards Ballyholme, laid out about 1905 on lands owned by Sir Daniel Dixon (hence its early name of *Dixon Avenue*), with most of the houses dating from the 1920s and put up by James Savage.
See *Lawrence 11224*.

No.15: Green Hall: *c.*1920: Two-storey red brick not very green house with leaded glass porch.

No.21: *c.*1930: Two-storey pebbledashed house with hipped rosemary-tiled roof, projecting dentilled eaves, and pebbledashed chimneys with brick tops.

No.23a: *c.*1925: Small bungalow with hipped roof and central chimney. With chunky quoins under overhanging eaves, dentilled arch forming porch to front door - lions crouching either side - and walls covered in broken coloured glass in fake ashlar bands.

Nos.22-24: *c.*1930: Two-storey roughcast semis with paired half-timbered gables and porches in outer bays with striped columns; notable for chalk pebble quoinstones with neatly-placed black pebble centres.

No.40: *c.*1930: Two-storey hipped roughcast house with battered ground floor cheeks to front elevation, and first floor windows immediately under eaves; stained glass porthole window, and tiles arranged in sunrise pattern over front door.

SHANDON PARK EAST

Steep cul-de-sac off Ward Avenue with houses set well back from the road, laid out in a T-shaped plan about 1925.

No.6: *c.*1930: Hipped bungalow with quoins and canted bays; massive palm tree in front.

No.22: *c.*1930: Two-storey roughcast house at the top looking over the park; double-fronted two-storey chequered gable set forward; leaded lights, rosemary-tiled porch with timber kneelers.

SHANDON PARK WEST
Cul-de-sac off Clifton Road, laid out during the 1920s on an attractive hilly site.

No.10: Rosedene: *c*.1934: Extraordinary villa with half-hipped pantile roof and knobbly ridge, central brick chimney; verandahs either side of central feature which rises to form a hipped dormer with quoins of red tile; windows with leaded casements. Trim little garden to match.

Shandon Terrace: See 11-29 Princetown Road.

SHERIDAN DRIVE
Although not developed till about 1900, this road from what is now Ballyholme Esplanade to Groomsport Road was present before 1830. Most of the present buildings on the west side date from about 1910, and most on the east side from the 1920s. The name presumably came from Helen Sheridan, the granddaughter of playwright Richard Brinsley Sheridan and mother of the 1st Marquess of Dufferin and Ava.

Ballyholme Roman Catholic Church: 1931, by E & J Byrne: Roughcast Romanesque barn church with cockscomb ridge, raised gable ends and crosses on finials. Gabled porch on side and simple railings. This was Edward Byrne's last work and he did not live to see its completion. Builders Thomas McKee & Sons, with leaded lights by Clokey and pews by Trevor C Thomas.
See *Crosbie p.20; IB 7 Jun 1930, 20 Jun, 1 and 15 Aug 1931; Northern Whig 27 Jul 1931; Spectator 1 Aug 1931; WAG 394.*

Nos.1-7: *c*.1890: Two-storey stucco terrace with canted ground floor bays; margin-paned windows to ground floor; stucco chimneys with tall octagonal pots.

SHORE STREET: See *Queen's Parade*.

SHREWSBURY DRIVE
Short road of pebbledashed terrace houses from Clandeboye Road to Chester Park, laid out during the 1930s.

SILVERSTREAM ROAD
Road parallel to and south of the Belfast Road, laid out by 1939 but developed after the war.
See *Spectator 24 May 1952.*

SIX ROAD ENDS
The hamlet of Six Road Ends is in the townland of *Ballygrainey* to the south of Bangor at the junction of the roads to Carrowdore, Newtownards, Clandeboye, Gransha, Ballycrochan and Donaghadee. Not surprisingly with modern traffic, it attracts very little pedestrian movement and has no sense of community left.

Ballygrainey Presbyterian Church: dated 1837: Two-storey hipped-roof smooth-rendered barn-plan church with quoins and simple roundheaded Gibbsian doorcase with spider-web fanlight on the gable towards the crossroads. Sides four bay with roundheaded fixed lights at first floor and double-hung sashes on ground floor, all small-pane. On the Gransha Road.

Betsy Gray's Cottage: The derelict rubble-stone cottage reputedly once lived in by the heroine of Lyttle's novel of the 1798 uprising, *Betsy Gray*, was uninhabited in 1940, and has been derelict for many years. It stands "in a little clearing among some very old and very tall trees" a short distance along the Carrowdore road. Betsy was a young girl who followed her brother and lover, both Presbyterian United Irishmen, to the Battle of Ballynahinch, where they were killed.
See *Robinson pp.78-79; Spectator 21 Sep 1940, 20 May 1950.*

SIXTH AVENUE
Narrow cul-de-sac off Fifth Avenue in the *Bay Lands* development, laid out about 1924. The houses are small and closely-packed, with the gable of no.7 a modest eyecatcher at the end of the lane.

SOMERSET AVENUE
From Grays Hill with an elbow down to Queen's Parade at the Marine Gardens, developed around 1880.
See *Lawrence I2685.*

Nos.1-7: *c.*1910: Terrace similar to Princetown Avenue just above, pebbledashed with red brick ground floors and consistent Arts and Crafts detailing. No.1 is gabled and three-storey, with a leaded light side porch, the remainder are two-storey with two-storey brick canted bays and dormers above.

Nos.2-16: Bowman's Cottages: *c.*1880: Charming terrace of two-storey stucco houses with ground floor bay windows and frilly bargeboards to cheek-by-jowl dormers; round-headed windows to first floor, twinned under alternate dormers. Very few houses altered until recently, when most have been roughcast, but no.14 remains original. Named after James Bowman, who built them.
See *Lawrence I2685.*

Nos.18-20: *c.*1895: Two-storey brick houses with basements, with gables set forward containing tiny paired blind lancets in gables.
See *Lawrence I2685.*

Nos.26-28: *c.*1880: Pair of two-storey stucco houses on angle of the street, with narrow horizontally divided sashes.
See *Lawrence I2685.*

Somerset Terrace: See 1-5 Princetown Road.

Bangor West Church, Rathmore Road: built in 1963, its architect W McK Davidson set out to design a memorable building, and it is still quite disconcerting today. (Peter O Marlow).

Telephone Exchange, Southwell Road: the 1935 exchange building is a symmetrical neo-Georgian design in brick and ashlar, with two rather jazzy terracotta plaques proclaiming the tyranny of the telephone. (Peter O Marlow).

37-47 Southwell Road: Maryville Crescent dates from about 1890, and retains its rhythm of gables and bay windows. Not exceptional in its day, the completeness of the terrace is now uncommon. (Peter O Marlow).

Somerset Avenue: this photograph taken about 1890 shows Bowman's Cottages on the right, and in the background Quay Street, Bridge Street and the piecemeal early development of Clifton Road. (Lawrence Collection).

SOOTY RAW: See *King Street*.

SOUTER'S ROW: See *King Street*.

SOUTHWELL ROAD

Formerly *Southwell Street*, this existed as a track in the mid 19th century, following the line of a stream that had run above ground up till about 1850, but only began to be developed in the 1880s, from Queen's Parade to Dufferin Avenue. Its name is derived from the *South Well*, now built over (possibly near the foot of King Street), where St Comgall reputedly cured a monk's blindness.
See *Milligan pp.18-19; NDH 22 Feb 1889; Seyers p.2*.

Nos.19-23: 1902, by Henry Chappell for William Hanna: Terrace of three-storey stucco houses with Dutch gables, very tightly curved corner to King Street and very narrow two-storey bow windows.
See *App 53*.

Nos.37-47: Maryville Crescent: *c*.1890, probably by John Neill: Two-and-a-half storey stucco terrace with two-storey canted bays and ornamental bargeboards to gables and dormers. Many houses still have double-hung sashes.
See *BNL 19 Oct 1892; NDH 22 Feb 1889*.

196

Telephone Exchange: 1935, by Richard Ingleby Smith of Ministry of Finance: The original telephone exchange is a Classical building in Flemish bond brickwork, three storeys tall with cornice and ground floor in sandstone ashlar; five-bay design with alternate niches at first floor whose symmetry is broken only by the ground floor door. Two terracotta plaques above the first floor windows depict wildly ringing telephones. The *National Telephone Co Exchange* had been at no.7 from 1908. The new extension alongside at **nos.49-65**, with its high forbidding railings preventing access to the gaps between its piloti, is of the "uncompromising" school.

195

Nos.14-20: Elsinore Terrace: *c*.1900: Three-storey terrace of stucco houses with semicircular-headed second floor windows and two storey canted bays.

Nos.22-30: *c*.1910: Terrace of ornate three-storey stucco houses, with scalloped parapets to two storey canted bays; windows segmental-headed with unusual eight-pane upper sashes.

Nos.34-36: Brookfield: *c*.1890: Pair of two-and-a-half storey painted brick houses with canted ground floor bay and ornamental bargeboard dormers. No.38 is slightly smaller but of similar date.

Nos.62-80: Royston Terrace: *c*.1900: Terrace of stepped two-storey red brick houses with scrolled dormers, many of whose ground floor bays are linked by closed verandahs to brick partitions.

Nos.82-86: Three-storey stucco houses with basements, linked to Landerville Crescent in Dufferin Avenue.

SPRINGFIELD AVENUE
Steep road bending down from Hamilton Road, following the line of the stream from Ward Park, then bending sharply into Springfield Road. The stream was dammed at this point (known as the *Dam Bottom*) to provide a head of water for the old corn mill in Mill Row; the street was developed about 1905 and the river was culverted over shortly after. A curious mixture of double-fronted stucco houses facing Ward Park and much smaller terrace houses at the bottom of the hill.
See *BHS II p.24*.

Nos.1-19: Riverside Terrace: *c*.1905: Stepped terrace of two-storey stucco houses with corbelled roofs to rectangular ground floor bays and chamfered openings.

No.2a: *c*.1905: Two-storey double-fronted stucco house; canted bays with terracotta finials flank a bowed cast iron balcony in the central bay. All windows unfortunately replaced.
See *Lawrence 11234*.

SPRINGFIELD ROAD
Street of mainly terraced houses from Prospect Road to Bingham Street, developed around 1910.
See *Wilson p.80*.

Nos.1-3: *c*.1910: Two-storey roughcast semis with projecting rafters at eaves.

Nos.5-7: Northdene: *c*.1907: Two-and-a-half storey red brick houses with twin finialed dormers and cockscomb ridge.

Nos.9-11: Glenroyd: *c*.1905: Pair of two-and-a-half storey stucco semis with canted ground floor bays; moulded surrounds to openings, with vegetable keystones at ground floor.

No.2: *c*.1908: Two-storey double-fronted villa with brick ground floor and rendered first floor; ornamental bargeboard to gables and dormers.

Nos.12-56: *c*.1900-1915: Terrace of stucco houses, most linked with running bays. Nos.54-56 were built about 1900, the remainder following about 1912.

SPRINGHILL ROAD
Winding road from the end of the Bryansburn Road uphill to the Belfast Road where it becomes the West Circular Road, developed after 1950 with rustic brick houses. SPRING HILL, a farmhouse which in 1833 had orchards at front and back, stood in the line of the road close to the junction with Lyndhurst Avenue, and was demolished in 1983.
See *Spectator 23 Oct 1983*.

SPRINGWELL DRIVE, Groomsport
Steep hill winding up from the junction of Main Street and Donaghadee Road,

past The Hill. Now leads to various housing estates, but was undeveloped before the war.

SPRINGWELL ROAD, Groomsport
Road from Groomsport to the upper Donaghadee Road out of Bangor, for the most part entirely rural with good hedgerows and grass verges.

Orange Hall: 1933: Simple roughcast gabled hall with foundation stones set into the wall on either side of the door. This replaced an earlier hall that stood on the Main Street.

No.2: c.1885: Stucco cottage with yellow brick quoins and window heads; sandstone skews and tall octagonal chimneys, which clearly associate it with the now Groomsport House Hotel. Shouldered gables.

STANLEY AVENUE
Short cul-de-sac off Stanley Road, laid out about 1930.

STANLEY ROAD
Road rising from Holborn Avenue to Clifton Road, laid out before 1903 but not developed till after 1920.

STATION DRIVE, Carnalea
Off north side of Killaire Avenue, mostly developed with chalets and semi-bungalows between the wars, now all redeveloped or altered.

STATION ROAD, Carnalea
Road from Crawfordsburn Road into Carnalea, developed before 1900.

Railway Bridge: c.1870: Coursed random sandstone single-arch bridge with rusticated voussoirs and blackstone battering.
See *McCutcheon pp.179-80.*

Carnalea Golf Club: Slate-hung building with "mansard" upper floor, on the site of the former *Royal Belfast Golf Club*, which had moved its links from Holywood to Carnalea in 1892. The course was "only a nine-hole one, but is very sporting", and there was a "comfortable and ornamental club-house for the members". When the Royal Belfast moved to Craigavad in 1926, Bangor Borough Council took the course over and extended it to eighteen holes.
See *Crosbie p.57; Lawrence 9531; Praeger p.67; Spectator 28 June 1930.*

STATION SQUARE, Carnalea
Accessed from Station Walk and Station View, relegated to a back route now that the car has replaced the railway as the main form of access to Carnalea.

Station House: c.1895: Two-storey red brick stationmaster's house with black string-courses, moulded brick cornice and remains of apex board.
See *McCutcheon p.179.*

STRICKLANDS

Here in the 1860s, where Bryan's Burn runs into Smelt Mill Bay (named possibly from a nearby lead smelting mill, or possibly from the combination of small fish (smelt) near the scutch mill at the top of the glen), lived an old lady called Sally McDonald who "gathered dulse, limpets and wilks... [she] claimed the shore and never paid rent or taxes".

A scutch mill formerly stood at the top entrance to the glen, but it was already marked as an "old mill" in 1833. The Glen was purchased in 1913 from Col Crawford, and after much debate about the respective virtues of clean parkland versus romantic wilderness, a rustic bridge was built in 1914 and about the same time on an area known as *Fairy Green*, a gabled and verandahed boy-scout type construction of timber logs known as THE BUNGALOW was built to provide restaurant facilities. Long-established residents complained about "the playing of musical instruments, the firing of revolvers, the use of obscene language in horseplay of various kinds", but the new facility was much used. The Bungalow was removed about 1940 and the Glen returned to wilderness. It remains little-spoilt, with twisting paths following the path of the burn at different levels.

See *Eakin; Lawrence 9549; Seyers pp.14, 52-53; Spectator 9 Jan and 26 June 1914; Wilson p.12.*

SUMMERHILL PARK

Roughly crescent-shaped road leading to a pedestrian link to Grove Park, partly developed by 1939.

T

TENNYSON AVENUE

Primarily a short road linking Princetown Road to Farnham Road, the name is also given to its continuation as a lane down to the sea. In existence before 1833, when it was known as the *Wrackey Loanen*, presumably because the sea wrack was brought up the lane to be spread on the fields; houses were built on it about 1890.
See *Seyers p.13.*

Nos.9-11: *c.*1900: Pair of two-and-a-half storey semi-detached houses with Dutch-gable dormers and two-storey canted bays; fluted pilasters to doorcases and to small dormer windows. In 1910, no.9 was listed as *Ivy Lawn*.

Nos.2-8: *c.*1890: Two pairs of stucco semi-villas with broad two-storey canted bays, and side porches. Windows in moulded surrounds with keystones; corbelled stucco chimneys. In 1910, nos.2 and 8 were named respectively *Fernbrook* and *Huntly*.

THE HILL, Groomsport

Terrace of houses running parallel to the Main Street but inland on the spine of a hill. Laid out in the first part of the 19th century, it was probably originally single-storey for the most part, but it is now two-storey, with most of the houses modernised. At the back is a lane, *Back Hill*, the portion at the Bangor end being known as *The West Hill*.
See *Crosbie p.42; Eakin; WAG 1295*.

No.33: Bay View: *c*.1880 and later: Single-storey stucco house almost overpowered by two large wallhead dormers, with vertically divided double-hung sashes and rusticated keystones to each ope.

No.37: *c*.1830 and later: Pleasant gloss-painted end of terrace house with round-headed first floor window in dormer, four-panel door and vertically-divided double-hung sashes. Interesting irregular small windows to rear elevation.

THE POINT, Groomsport

Cul-de-sac at the west end of Groomsport Main Street, wrapping round the bay and leading to the path round *Ballymacormick Point*. Only a short walk from the twee benches and car parks of the harbour, this is a natural pebbled beach where oyster catchers and herring gulls wheel over ox-eye daisies and lichen-covered rocks.
See *Lawrence 10194*.

No.2: *c*.1935: Two-storey roughcast house with central gable set forward with shingled apex over half-timbering.

The Watch House: possibly 18th century: Two-storey roughcast house on peninsula enclosing the bay; hipped roof and meandering tall chimney-stack; canted bay to NE end, with a fine view out to sea. Formerly the *Coast Guard Station*, whose staff occupied a row of small cottages nearby that were demolished *c*.1965, its former slipway through the rocks can still be traced. In 1821 a lease of land was made from nearby Islet Hill Farm to the Commissioners of Custom to permit the "Preventative Water Guard" to operate their dwellings and boat house on this location, and it is possible that the building dates from that time.
See *Crosbie p.39; Eakin; information from Denis Mayne; WAG 3134*.

THIRD AVENUE

Short street from Donaghadee Road leading into the *Bay Lands* development at Fifth Avenue, commenced in 1925.

No.6: *c*.1930: Red brick bungalow with hipped slate roof, terracotta finials and gable chimneys. Canted bays; stained glass toplights on either side of half-glazed door. On raised site with laurel hedge and copper beech.

THORNLEIGH GARDENS
Road rising uphill from Donaghadee Road, laid out at the end of the 1930s but mostly developed post-war. The bottom of the road had previously been used as a BRICK WORKS, with the factory roughly on the site of the present shopping centre, kilns on the other side of the road and a clay pit on the present Grove Hill park across the Donaghadee Road.

Tower Buildings: See 2-8 Victoria Road.

TOWER ROAD, Conlig
Road from Main Street in Conlig to Clandeboye Golf Club.

No.3: Two-storey stucco house with narrow roll mouldings to window reveals, and off-centre porch; low boundary wall overgrown with ivy.

U

UNION STREET: See *Holborn Avenue*.

Upper Clifton: See Carrisbrooke Terrace and Victoria Terrace.

V

VALENTINE ROAD
Road off Castle Park Road leading to the back of Bangor Castle and Valentine playing fields.

Home Farm: *c.*1850: Quadrangle of rubble-stone buildings, including two houses, with dressed sandstone quoins and brick opes. Nearby is the brick *walled garden*, also associated with Bangor Castle.

VENNEL, THE
Narrow entry from Queen's Parade to King Street, originally containing half a dozen houses, mostly demolished in recent years.

Verington Terrace: See 35-53 Prospect Road.

VICTORIA DRIVE
Short elbow off Victoria Road, developed in the 1920s.

No.38: *c.*1930: Pebbledashed double-fronted hipped bungalow with scalloped red slates and stepped gable chimneys; stained glass toplights. **Nos.30-34** are similar but altered.

2-8 Victoria Road: the Tower Buildings of about 1890 get their name from the adjacent Custom House but are distinctive in themselves with their oriel windows on timber kneelers. (Peter O Marlow).

36-38 Victoria Road: the broad bay window on no.36 marks the former Royal Irish Constabulary barracks; the station sergeant lived in the neighbouring house. (Peter O Marlow).

The Bungalow, Stricklands Glen: built about 1915 to provide restaurant facilities in the newly acquired Glen, the rustic building was cleared about 1940. (From an old postcard).

9-11 Ward Avenue: designed in 1900, this pleasant pair of unpainted stucco houses has changed little apart from insertion of plastic windows in one house. (Peter O Marlow).

VICTORIA ROAD

Originally called *Fisher's Hill*, probably because the inhabitants of many of the former single-storey whitewashed houses along the road in the early 19th century made their livelihood from the sea. By 1850 many were carpenters, including one Captain Rippett. The lower part was known as *Corporation Street* in 1833. After King Edward VII's visit the road was renamed; it had already been largely rebuilt in the previous decades. Bangor's first Presbyterian CHURCH stood here about 1650.
See *Seyers pp.8, 12.*

Nos.57-65: Wavemount Terrace: *c.*1900: Terrace of two-and-a-half storey stucco houses with bargeboards to wallhead dormers and label mouldings with bosses to ground-floor windows.

Nos.121-125: Clifton Mount: *c.*1900: Terrace of two-and-a-half storey stucco houses with canted bays at ground floor, and until recently skews and ball finials to dormers.

Nos.2-8: Tower Buildings: *c.*1890: Terrace of two-and-a-half storey stucco houses with canted oriels at first floor supported on timber brackets; gablets with apex boards and ball finials; paired doorcases with heavy corbels; margined panes. Built by Charles Neill.
See *Seyers p.12; Welch 13; Wilson p.49.*

Nos.10-14: *c.*1990: Three-storey rendered block with metal feature at second floor. The terrace previously on this site included probably the oldest HOUSES in the street, two-storey houses of *c.*1860, with small broad-framed double-hung sashes.

Nos.16-26: *c.*1900: Terrace of two-and-a-half storey stepped stucco houses with ornamental bargeboards to dormers; remains of shopfront at no.26 complete with corbels, pilasters and double door, and render stippled to represent ashlar.

Nos.28-34 and 40-48: Westminster Terrace: *c.*1895: Terraces of two-and-a-half storey stucco houses with two-storey canted bays, apex boards to dormers, moulded dentilled string course at first floor and five-panel doors.

Nos.36-38: *c.*1900: Terrace of two-storey stucco houses with ground floor canted bays, label moulding at first floor, and dentilled cornice. Unusual fenestration, with upper sashes divided into three and lower into two, with mostly original glass. In the early years of the century, no.36 was the *RIC barracks*, and its neighbour was occupied by the station sergeant and his family.

Nos.52-72: *c.*1900: Terrace of two-and-a-half storey stucco houses, with ornamental bargeboard to wallhead gable, and continuous cill course to first floor windows. Nos.62-72 have ball finials to dormers.

Nos.118-126: *c.*1910: Terrace of two-and-a-half storey stucco houses with two-storey canted bays; fielded four-panel front doors.

VICTORIA TERRACE

Terrace of houses alongside Carisbrooke Terrace, reached by a lane off Clifton Road. This may originally have been called *Ardbraccan Terrace*. Nos.32-42 Ballyholme Esplanade were also called Victoria Terrace.

Nos.1-7: *c.*1875-80: Terrace of two-and-a-half storey stucco houses with two-storey canted bays, channelled ground floor, with alternating forms of dormer. Rolling lawns terrace down towards the sea from this commanding site above Seacliff Road. Built in stages from the Upper Clifton end.
See *Hogg 59; Lawrence C5054; Welch 20, 21; Wilson pp.29, 31.*

VIMY RIDGE: See *Central Street*.

W

WARD AVENUE

Long road from Donaghadee Road linking to Clifton Road and on down to the sea at Luke's Point. Although much of it was in existence in 1833, and it was completed (and known as *Bachelor's Walk*) before 1858 to serve the Ward Villas and take a track down to the sea, it was at the end of the 19th century that building development started in earnest.
See *BHS II p.24.*

No.7: *c.*1910: Two-storey double-fronted stucco house with two-storey bays and fishscale-slated aedicule over front door.

Nos.9-11: Ambleside: 1900, for Robert Brown: Handsome pair of two-and-a-half storey double-fronted semi-villas with outer bays canted and inner ones rectangular; ornamental bargeboards to gables and pitched-roof dormers, fielded panel doors, conservatories to rear.
See *App 12.*

Nos.15-25: *c.*1900: Two-storey double-fronted detached stucco houses with segmental-headed sash windows, and castellated bow windows at ground floor; of which only no.17 is now intact. No.21 has been recently demolished and replaced with a modern house.

Nos.27-29: *c.*1920: Pair of roughcast semis with cement trims to opes, and Arts and Crafts windows (altered in no.27); entrance porch at side. Possibly *St Malo*, designed for Jacob O'Neill by Thomas Callender.
See *IB 28 Oct 1916 p.536.*

No.31: Fintimara: *c.*1905: Two-storey double-fronted stucco house with plain sashes, glass canopy over door with glazed light, sidelights and fanlight.

No.53: *c.*1910: Two-storey double-fronted red brick house with roughcast first floor and half-timbered gables projecting above canted bay windows.
No.51 is similar in feeling.

No.8: *c.*1905: Two-storey double-fronted stucco house with two-storey canted bays; rather good stained glass windows.

No.10: Alexandraville: 1902, by Henry T Fulton: Two-storey double-fronted stucco house with two-storey bow windows.
See *App 45.*

Nos.14-16: 1900, by J Fraser & Son: Two-storey semi-detached houses in stucco, with two-storey bow windows and common glazed porch.
See *App 4.*

No.34: Waihi: *c.*1905: Double-fronted unpainted stucco house with chamfered opes and dentilled cornice.

No.42: *c.*1910: Trim bungalow with bellcast blue slate roof and turrets (one octagonal, the other square) at each end of front elevation, which is partly verandahed.

Nos.44-46: *c.*1910: Pair of detached houses, gabled to one side, hipped to other, with wavy bargeboards and bulgy battered chimney-stacks in rosemary-tiled roof. Doors inset in brick arched openings, with stained glass lights.

No.48: Erindale: *c.*1910: Two-storey double-fronted stucco house with two-storey canted bays and ornamental bargeboard.

Nos.82-84: Baymount: *c.*1870: Pair of substantial semi-villas overlooking Ballyholme Bay, two storeys in height with basements at rear; walls pebbledashed, some windows margin-paned. Was a preparatory *school* in the 1930s, but built originally as a residence for Rev Stewart.
See *Seyers p.11.*

Ward Park: See Hamilton Road.

Wavemount Terrace: See 57-65 Victoria Road.

WAVERLEY DRIVE

A pleasant mixture of two-storey houses in various 20th century styles - the road had been laid out before 1903, and development was completed about 1930.

No.1: *c.*1900: One-and-a-half storey detached three bay stucco house with quoins; shallow oriels at ground floor, three frilly bargeboard dormers at wallhead. Windows unfortunately all plastic.

Nos.5-7: Waverley Cottages: *c.*1890: Terrace of two-storey stucco houses with frilly bargeboards to dormers; nos.5-7 low, **no.9** (*Waverley House*) much taller, originally with a side conservatory.

Nos.35-41: *c.*1925: Two pairs of roughcast two-storey semis with narrow-pane upper sashes over plain lower sashes, and outer first floor windows Palladian with leaded lights. **Nos.38-40** are similar.

No.8: *c.*1905: Two-storey stucco house with timber verandah at one corner, round-headed windows at first floor with bossed hood mouldings, bracketed eaves.

WELL ROAD: See *Holborn Avenue.*

West End Terrace: See 98-108 Dufferin Avenue.

Westminster Terrace: See 28-48 Victoria Road.

WEST PLACE: See *King's Place.*

WEST STREET: See *King Street.*

WESTWARD HO!: See *Kensington Park.*

WILLIAMSON'S LANE: See *Fairfield Road.*

WINDMILL LANE
Lane linking Windmill Road to Bellevue, laid out about 1930.

WINDMILL ROAD
Road from Groomsport Road uphill to the Donaghadee Road, present before 1833, when a WATER MILL and its mill race occupied much of the eastern side of the road. The mill appears to have been in use as late as 1921, and it latterly made bone manure. It was "disused" by 1932, and being developed by 1939. There was also a water-driven threshing machine in use at the turn of the century. Until cut off by the development of the parallel *Bellevue*, the Ballyholme windmill was accessed to the west of this road.
See *Seyers p.12.*

Nos.1-15: *c.*1925: Four pairs of two-storey hipped roughcast semis with timber apex boards and finials to gables; depressed arch recesses with Moorish keystones. No.3 has duple sash windows, with coloured margin panes; no.11 is also well preserved, with double-hung sashes and elegant etched glass door, and a bossy brass plate reading "Ring the Bell".

Nos.8-14: *c.*1935: Two pairs of semi-detached smooth-rendered houses with Art Deco curved bay windows, originally with horizontal glazing bars and curved glass at the corners; all now with clumsy plastic windows.

WINDSOR AVENUE
Road laid out off the Bryansburn Road about 1900. Essentially a cul-de-sac, but Windsor Park leads off at the bottom.

Nos.1-9: *c.*1905, probably by John Russell for James Neill: Five detached roughcast houses, mostly unpainted, roughcast with string courses at first floor, and ground floor canted bays.
See *App 91.*

No.11: *c.*1900, probably by J C McCandless for William Reid: Two-storey double-fronted stucco house with two-storey canted bays; pierced red clay ridge, stucco chimneys, ornamental bargeboard to gable with finial and unusual apex board. Basket-headed doorcase with fanlight and sidelights.
See *App 24.*

No.2: *c.*1900: Low two-storey roughcast house, double-fronted with shallow rectangular ground floor bays and dormers to segmental-headed windows over.

No.14: *c.*1930: Large stucco house with yellow brick chimneys. Canted bays extend to form porch over central door, with central dormer above breaking roofline. Stained glass upper lights.

WINDSOR GARDENS
Steep road from Manse Road down to a pedestrian lane leading to Railway View Street, planned around 1928 and fully developed by 1939, mostly with stepped terraces.

WINDSOR PARK
Narrow road from Manse Road to Windsor Avenue, laid out around 1930 on ground that had been used as allotments in the early years of the century.

Winifred Terrace: See 4-12 Gray's Hill.

WOODGREEN
Cul-de-sac off Belfast Road, laid out about 1925. Redeveloped for Bangor Borough Council 1970-71, with a line of cherry trees, by H A Patton.

WRACKEY LOANEN: Former name for *Tennyson Avenue.*

> FOR GOD SO LOVED THE WORLD
> THAT HE GAVE HIS ONLY BEGOTTEN SON
> THAT WHOSOEVER BELIEVETH IN HIM
> SHOULD NOT PERISH BUT HAVE
> EVERLASTING LIFE

BIBLIOGRAPHY

Adamson: Adamson, Ian, *Bangor, Light of the World*. Bangor, Pretani Press, 1979.

Allen: Allen, Harry, *Town Trails 1 and 2 for Bangor*. Bangor, Seacourt Teachers' Centre, 1978.

App: Bye-law applications made to Bangor Urban Council 1899-1972 (previously held at the Planning Office in Downpatrick).

Arch Surv: *An Archaeological Survey of Co Down*. Belfast, HMSO, 1966.

Baddeley: Baddeley, M J B, *Ireland (Northern Counties)*. London, Nelson, 1909.

BAR: *Buildings At Risk*. Belfast, UAHS, annual reports 1994 to present.

Bassett: Bassett, George Henry, *County Down Guide and Directory*. Dublin, Sealy Bryars & Walker, 1886 (reprinted Belfast, Friar's Bush Press, 1988).

BHS: Bangor Historical Society, *Journal*.

Bingham: Bingham, Madeleine, *Peers and Plebs: Two Families in a Changing World*. London, Allen & Unwin, 1975.

BNL: *Belfast News Letter*.

Blitz: [Anon, introduction by Christopher D McGimpsey], *Bombs on Belfast; The Blitz 1941*. Belfast, Belfast Telegraph, 1941 (reprinted Bangor, Pretani Press, 1984).

BT: *Belfast Telegraph*.

Boucher: Boucher, Shelagh, *The Savoy Hotel*. Unpublished dissertation for Open University, 1980.

Brett: Brett, C E B, *Court Houses and Market Houses of the Province of Ulster*. Belfast, UAHS, 1973.

BT Guide: Belfast Telegraph, *Guide to Belfast and Surrounding Districts*. Belfast, Baird, 1934.

Camblin: Camblin, Gilbert, *The Town in Ulster*. Belfast, Mullan & Son Publishers Ltd, 1951.

Clandeboye: Rankin, Peter (ed), *Clandeboye*. Belfast, UAHS, 1985.

Country Club: *Crawfordsburn Country Club Diamond Anniversary 1937-1997*. Crawfordsburn, 1997.

Crosbie: Crosbie, Jane E M, *A Tour of North Down 1895-1925*. Belfast, Friar's Bush Press, 1989.

Dean: Dean, J A K, *The Gate Lodges of Ulster*. Belfast, UAHS, 1994.

Dixon: Dixon, Hugh, *An Introduction to Ulster Architecture*. Belfast, UAHS, 1975.

Dubourdieu: Dubourdieu, Rev John, *Statistical Survey of the County of Down*. Dublin, Graisberry and Campbell, 1802.

Eakin: Eakin, Terry, and Purvis, Ronnie, *Postcard Recollections of Bangor*. Holywood nd [1985].

Ewart: Ewart, L M, *Handbook of the United Diocese of Down & Connor & Dromore*. Belfast, Marcus Ward & Co, 1886.

Gallagher: Gallagher, Lyn, and Rogers, Dick, *Castle, Coast and Cottage: The National Trust in Northern Ireland*. Belfast, Blackstaff Press, 1986.

BBLIOGRAPHY

Girouard: Girouard, Mark, *The Victorian Country House*. New Haven, Yale University Press, 1979.

Green: Green, E R R, *The Industrial Archaeology of Co Down*. Belfast, HMSO, 1963.

Haire: Haire, Rev Robert, *A Short History of Wesley Centenary Methodist Church and Early Methodism in the Bangor Area*. Bangor, 1966.

Hall: Hall, Mr & Mrs S C, *Ireland, Its Scenery, Character etc*. London, Hall & Virtue, 1841-43.

Hamilton: Hamilton, Rev James, *Bangor Abbey Through Fourteen Centuries*. Belfast, John Aiken & Son, nd.

Handbook: *Handbook of the United Diocese of Down Connor & Dromore*.

Hanna: Three pictures of Queen's Parade and Quay Street by Thomas Hanna about 1860 - in the collection of North Down Heritage Centre.

Harris: Harris, W, *The Antient and Present State of the County of Down*. Dublin, 1744; reprinted by Davidson Books, Ballynahinch, nd.

Hogg: The Hogg collection of photographs is held in the Ulster Museum: unless otherwise stated, the references carry the prefix H05/15/ - for example the full reference for Hogg 32 is H05/15/32.

Hewitt: Hewitt, John, *Kites in Spring*. Belfast, Blackstaff Press, 1980. Hewitt's autobiographical poems are also reprinted in his *Collected Poems*, Blackstaff Press 1991.

IB: *Irish Builder*.

Ir Arch Arch: Irish Architectural Archive, Dublin.

Kirkpatrick: Kirkpatrick, Noel, *Take A Second Look (around Co Down)*. Alkon Press, Newtownards, 1993.

Knox: Knox, A, *A History of the County of Down*. Dublin, Hodges, Foster, 1875.

Lawrence: Lawrence Collection of photographs (mostly 1880-1915), National Library of Ireland (available on microfilm at Belfast Central Library).

Lewis: Lewis, S, *Topographical Dictionary of Ireland*. London, S Lewis, 1837.

Lowry: Lowry, T K, *The Hamilton Manuscripts*. Belfast, Archer & Sons, 1867.

Luckombe: Luckombe, Philip, *A Tour Through Ireland in 1779*. Dublin, Whitestone, 1780.

Lyttle: Lyttle, W G, *The Bangor Season*. Bangor, W G Lyttle, 1885 (reprinted Belfast, Appletree Press, 1976).

McCutcheon: McCutcheon, W A, *The Industrial Archaeology of Northern Ireland*. Belfast, HMSO, 1980.

Maguire: Maguire, Dean Edward, *Fifty Eight Years of Clerical Life in the Church of Ireland*. Dublin, Church of Ireland, 1904.

Merrick: Merrick, A, *Gravestone Inscriptions, Co Down vol.17 Barony of Ardes*. Belfast, Ulster Historical Foundation, 1978.

Milligan: Milligan, Charles, *My Bangor*. Bangor, Spectator Newspapers, 1975.

Milligan 2: Milligan, Charles, *Second Thoughts*. Bangor, Spectator Newspapers, nd [*c*.1977].

Minutes: Minutes of Town Comissioners 1865-85 held in North Down Heritage Centre.

Molloy: Molloy, J, and Proctor, E K, *Belfast Scenery in Thirty Views*. Belfast, Proctor Morgan Jellen, 1832 (reprinted by the Linen Hall Library Belfast 1983).

BIBLIOGRAPHY

Monuments Record: The archive of Environment & Heritage Service at Hill Street Belfast contains a wealth of information of archaeological and industrial remains, along with more modest records of buildings.

Morton: Morton, Grenfell (ed.), *Victorian Bangor: An Essay in Local History*. Belfast, WEA and QUB Extra-mural Studies Dept, 1972.

Nelson: Nelson, Walter, *Historical Sketch of Groomsport Presbyterian Church*.

Nicolson: Nicolson, Harold, *Helen's Tower*. London, Constable, 1937.

NDH: *North Down Herald and Bangor Gazette*. (Cuttings books 1886-94 held in North Down Heritage Centre).

NDHC: North Down Heritage Centre has a useful collection of photographs of early Bangor, not so far catalogued.

O'Laverty: O'Laverty, Rev J, *An historical account of the Diocese of Down & Connor*. Belfast, 1878-84.

Old Custom House: North Down Borough Council, *History of the Old Custom House and Tower House*. Bangor, NDBC nd [*c*.1985].

OS: Ordnance Survey maps (surveyed 1833, 1858, 1901, 1919, 1926 and 1939).

OS Mems: Day, Angélique and McWilliams, Patrick (editors), *Ordnance Survey Memoirs of Ireland: Parishes of County Down II, 1832-4, 1837: North Down and The Ards*: Belfast, Institute of Irish Studies, Queen's University of Belfast, 1991; also on microfilm at PRONI.

Parl Gaz: *The Parliamentary Gazetteer of Ireland*. Dublin, A Fullarton & Co, 1844.

Patterson: Patterson, E M, *The Belfast and Co Down Railway*. Newton Abbot, David & Charles, 1982.

Patton: Patton, H A, *An Outline Survey of Bangor*. Unpublished dissertation for Edinburgh University, 1947.

Patton M: Patton, Marcus, *Historic Buildings in Bangor and Groomsport*. Belfast, UAHS, 1984 (this earlier edition of the present book contained some different photographs as well as text).

Pike: Pike, W T, *Ulster: Contemporary Biographies*. W T Pike & Co, Brighton, 1909.

Praeger: Praeger, Robert Lloyd, *Official Guide to Co Down, Belfast & Co Down Railway Co*. Belfast, Marcus Ward & Co, 1898.

Presb Hist: *A History of Congregations in the Presbyterian Church in Ireland 1610-1982*. Belfast, Presbyterian Historical Society of Ireland, 1982.

PRONI: Public Record Office of Northern Ireland.

Raven: Raven maps, now held in North Down Heritage Centre.

Reid: Reid, Tom, *Trinity Presbyterian Church, Bangor*. Bangor, Trinity Presbyterian Church, 1989.

Reeves: Reeves, William, *Ecclesiastical Antiquities of Down & Connor & Dromore*. Dublin 1847.

Robinson: Robinson, Kenneth, *North Down and Ards in 1798*, Bangor, North Down Heritage Centre, 1998.

Seyers: Seyers, William Charles, *Reminiscences of Old Bangor*. Belfast, Lion Gate Press, 1983. (Reprinted from Co Down Spectator 1932-33, when it was published in instalments.)

SBP: *South Belfast Post*.

Scott: Scott, Rev Charles, *The Abbey Church of Bangor*. Belfast, W & G Baird, 1882.

BIBLIOGRAPHY

Smith: Smith, Charles, *The antient and present state of the county of Down*. Dublin, A Reilly, 1744.

Spectator: *County Down Spectator*, 1904 to present. (Available on microfilm at Bangor Library; an *Index to the Co Down Spectator 1904-64* has been published by Jack McCoy for the SEELB, 1983.)

Spect YB: *Spectator Year Books*, 1906-71.

Stevenson: Stevenson, J, *Two Centuries of Life in Down, 1600-1800*. Belfast, McCaw Stevenson & Orr, 1920 (reprinted White Row Press, Belfast, 1990).

UJA: *Ulster Journal of Archaeology*.

Ulster Bank: Knox, William J, *Decades of the Ulster Bank 1836-1964*. Belfast, Ulster Bank, 1965.

VAL: Valuation Records in PRONI.

WAG: W A Green collection of photographs (available in Ulster Folk & Transport Museum).

Welch: Welch collection of photographs held in the Ulster Museum: unless otherwise stated, the references carry the prefix W05/15/ - for example the full reference for Welch 32 is W05/15/32.

Wilson: Wilson, Ian, *Bangor: Historic photographs of the County Down town 1870-1914*. Belfast, Friar's Bush Press, 1992.

Wilson Wm: Wilson, William, *350th Anniversary of 1st Bangor Presbyterian Church*. Bangor, 1973.

Young: Young, Robert M, *Belfast and the Province of Ulster in the 20th Century*. Brighton, W T Pike, 1909.

The Author

Marcus Patton was brought up in Bangor and studied architecture at Queen's University. He works for Hearth, a housing association managed by the National Trust and the Ulster Architectural Heritage Society to rescue historic buildings at risk in the province; and he is currently on the committee of the UAHS. In addition to writing the Society's *Central Belfast: An Historical Gazetteer* (1993) and *The Diamond As Big As A Square* (1981, in collaboration with David Evans), he has illustrated several guidebooks, including the Cadogan Press' *Ireland*, for which he travelled throughout the thirty-two counties. He is an Associate of the Royal Ulster Academy, and a compulsive musician. He was awarded the OBE in 1995 for services to conservation.

INDEX

(Numbers in bold refer to illustrations)

Abbey vii, 1, **5**
Assembly Room 122
Beard, restraint of whilst riding 183
Beaumont, Ivor 100
Belfast roof truss 6, 23
Bell, Mr 168, **171**
Betsy Gray, 194
Bingham, Hon Barry 96
Black Hole, The 131
Blair, Robert 132
Boyd, D W 12, 38
Boyd Partnership 38, 62
Braid, James 28
Brown, Capt Inkerman 45, 72
Browne, George 144
Burn, William 20, **34**, 35, 46, 48, 49, **87**
Butterfield, William 39
Byrne, E & J 69, 144, 193
Cage of Death 170
Callender, Thomas 130, 166, 206
Caproni's 190
Carnegie, Andrew 96
Castle 33-37, **34**, 202
Chalets 66, **214**
Chappell, Henry 17, 197
Cholera House 23
Christians, lack of 58
Cinemas 97, **99**, 120, 170
Close, S P 120, 159
Coal dust, perils of 170
Cook, Alan 49
Cooper, W 13
Cotton mills ix, 4, 103, 167
Craig, Vincent **ii**, 52, **53**, 59, 77, 112, **113**
Custom House **172**, 173
Cyclist, an embarrassing 187

Davidson, Peter 86
Davidson, Sir Samuel 115, 165
Davidson, W McK 183, **195**
Devon, Stanley 42, 55
Dispensary 74
Douglass, William 58
Dufferin, Lord 22, 46, **47**, 49, 52, **87**, 92
Eves, Fred 71
Falloon, Paddy 138
Ferguson & McIlveen 92
Ferrey, Benjamin 46, 49
Fitzsimons, Nicholas 85
Flappers, object lesson to 162
Fraser, J & Son 207
Fraser, James 48
Frazer, J 77
Fulton, Henry T 13, 31, 207
Garden gnomes 74
Gas Works 23, 108, **110**
Gaw, J A 91
Gifford & Cairns 149
Golf 97, 199, **216**
Gospel in a nutshell 22, **210**
Greenwood, Cecil 108
Hall Black Douglas 43
Halpin, George 58
Hamilton, James **164**, 165
Hamilton, Sir James viii, 2, 33, 152, 174
Hanna, Denis O'D 66
Hanna, James 18, 66, 121, **123**
Hanna, Jeannie 45
Hewitt, John 14
Hill, R Sharpe 89
Hobson, H A 28
Hodgins, L H 91, 173, 179

Inkpot, The 148
Jordan, Sir John Newell 8
Kennedy & Hill 187
Kerr, William 62
Kirk McClure & Morton 154
Knox & Markwell 97, 116
Lamont, E P 7
Lang, F W 167
Lanyon, Charles 65, **67**
Lanyon Lynn & Lanyon 6
Leith, Raymond 103, 108
Lime kilns 23
Lindsay, J G 14, 170
Lipton, Sir Thomas 55, 166
Love, Castor J **15**, 42
Lynn, W H 39, 46
Lyttle, W G 4, 12, 132, 194
Mack, Rev Isaac 75, 139
Macklin, T Eyre 96
Macrae, Alexander 121
Marine Pagoda 141
Market 39
Market House 122, **124**
McAdam Design 36, 142
McAlister Armstrong Partnership 159
McCandless, J C 77, 161, 209
McIlveen, Samuel 3
McKnight, Gordon 90, 91
Mendicity Institute 4
Methodists, rigorous 90
Millar & Symes **128**, 130
Minshull, W H 13
Moloney, Mr 57

215

INDEX

Mountjoy 153
Nash, John **60**, 61
Neill, Charles 7, 107
Neill, John 197
Neill, John McBride
 97, **99**, 100, 126
Neill, Robert 72, 179
O'Donnell & Tuomey 50
O'Neill, Gordon
 19, 78, **79**, 81, **83**,
 124, 125, 126
O'Neill, Jacob 43, 206
O'Neill, John 98
O'Shea, J J 92, 95
Ostick & Williams 119
Patteson, Rev William
 62, 141
Patton, H A
 28, 55, 56, 70, 125,
 131, 149, 187, 190, 209
Patton, Rev Alexander
 132
Pavilion of Varieties 6
Philips, J P 90
Phillips, J J & Son 181
Pickie 142
Poor House 44
Potatoes, underwater 10
Pound 4
Priestley, L A M 146, 187
Railway
 xii, **5**, 6, 28, 199
Raven, Thomas ix
Reavey, Sammy 44
Rippingham, T F O
 126, **127**
Robinson & McIlwaine
 125, 134
Robinson, Henry Lynch
 80, 81
Robinson Patterson
 Partnership 25, 100
Roome, W J W
 28, 70, 91
Rosie's Lumps 27
Russell, John 16, 208
Salvin, Anthony
 4, **34**, 35, 38
Sands, James 66, **68**
Savage, James
 12, 13, 92, 192
Savage, Robert N 100
Scheemaker, Peter 3
Schools xii, 4, **15**, 29,
 30, 38, 42, **47**, 49,
 54, 55, 56, 70, 84,
 87, 96, 97, 122, 131,
 135, 140, 155, 182
Seaver, Henry 182
Servant girl, distressed
 160
Shanks & Leighton
 128, 130
Simpson, Hugo 38
Smith, Richard Ingleby
 195, 197
Snow, use as drawing
 board 183
Somme 49
St Comgall vii, 1
St Malachy viii, 1, 4
Steamers 70, 153, xii
Stevenson, Samuel
 27, 144, 179
Swimming xiii, 10, 142
Taggart, W D R 144
Taggart, WDR & RT
 50, 84, 89
Tax avoidance 200
Temperance, attempted
 170, 176
Thompson, Larry 174
Thompson, Mortimer 29
Tomlinson, J 50
Tonic, The 97, **99**
Turnip, raw 98
Tyre, non-puncturable
 109
Walker, William 35
Walshe, William 109
War damage 7, 147
Ward, R E 4, 35
Water mills 10, 31,
 136, 146, 200, 208
Watt & Tulloch 59
Watt, Jenny 143
White, Derrick 169
Wilson, David 103
Windmills 22, **25**, 61
Woodgate, Robert 45, **47**
Woods, Ernest L
 9, 45, **54**, 56, 70, **94**,
 95, 96, 97, 144, **145**,
 162, 191, **216**
Wright, A J 52
Yachting
 ii, xiii, 52, **53**, 187
Young & Mackenzie
 40, 52, 56, 70,
 92, 117